PRAISE FOR
THE PSYCHOLOGY OF TOTALITARIANISM

"As I walk through the halls of a major US medical center, I see eyes that divert themselves away from me as I pass. When we engage in our usual discussions on patients, the topic of COVID-19 vaccination brings a halting response: 'We don't want to talk about it.' I see fear, shame, and a never-ending cycle of groupthink that has been more contagious among physicians than aerosolized SARS-CoV-2 in a crowded elevator. Mattias Desmet, like a guided missile, has hit the target. The medical community is in mass formation and this led to a much larger penumbra that has enveloped the general population. In this book, Desmet has constructed an explanatory framework from which the cohesive fabric is suspended that clearly and concisely explains what is happening and what the next steps are that each and every one of us need to take to break the 'spell' and restore normalcy. A must read for our time."

—PETER A. McCULLOUGH, MD, MPH;
chief medical advisor, Truth for Health Foundation

"Transcending medical controversies, this book offers an indispensable window into the social phenomenon we call COVID."

—CHARLES EISENSTEIN, author of
Sacred Economics and *The Coronation*

"Mattias Desmet is the world's expert on the phenomenon of mass formation—and one of the most sincere, thoughtful, and important intellectuals of the twenty-first century. If you want to understand why and how the coronavirus pandemic response unfolded the way it did at a societal level and—even more importantly—how to prevent such a travesty from happening again, *The Psychology of Totalitarianism* is essential reading. Desmet shows us how to reclaim our humanity in an increasingly dehumanized and mechanized world."

—DR. REINER FUELLMICH, trial attorney; cofounder, Berlin's Corona Investigative Committee

"In this masterful book, Desmet asks how we have arrived at the doorstep of totalitarianism. Taking the reader on a wild, scholarly ride through history, science, and psychology, he delivers answers both necessary and unexpected."

—HEATHER HEYING, PhD, evolutionary biologist; coauthor of *A Hunter-Gatherer's Guide to the 21st Century*

"Desmet is waking a lot of people up to the dangerous place we are now with a brilliant distillation of how we ended up here."

—ROBERT F. KENNEDY, JR.

"Mattias Desmet's theory of mass formation is the most important lens through which we can understand the COVID-19 pandemic and the social aberrations that accompanied it. In *The Psychology of Totalitarianism*, Desmet explains how and why people will willingly give up their freedom, how the masses can give rise to a totalitarian leader, and—most importantly—how we can resist these phenomena and maintain our common humanity. This is the most important book of 2022."

—DR. ROBERT MALONE, author of *Lies My Gov't Told Me*

"Mattias Desmet's [theory of mass formation hypnosis] is great.... Once I kind of started to look for it, I saw it everywhere."

—ERIC CLAPTON

THE PSYCHOLOGY OF TOTALITARIANISM

THE PSYCHOLOGY OF TOTALITARIANISM

MATTIAS DESMET

TRANSLATED BY ELS VANBRABANT

Chelsea Green Publishing
White River Junction, Vermont
London, UK

Originally published in Belgium by Pelckmans Publishers in 2022
as De Psychologie van Totalitarisme.

This edition published by Chelsea Green Publishing, 2022.

Translator: Els Vanbrabant
Translation Proofreader: Tineke De Cock
Project Manager: Patricia Stone
Developmental Editor: Brianne Goodspeed
Copy Editor: Angela Boyle
Proofreader: Nancy A. Crompton
Indexer: Shana Milkie
Designer: Melissa Jacobson

Printed in the United States of America.
First printing May 2022.
10 9 8 7 6 5 4 3   22 23 24 25 26

ISBN 978-1-64502-172-8 (hardcover) | ISBN 978-1-64502-173-5 (ebook)
| ISBN 978-1-64502-174-2 (audio book)

Library of Congress Cataloging-in-Publication Data is available upon request.

Chelsea Green Publishing
85 North Main Street, Suite 120
White River Junction, Vermont USA

Somerset House
London, UK

www.chelseagreen.com

CONTENTS

Introduction

To write a book about totalitarianism—the idea first occurred to me on November 4, 2017. Or rather, it first appeared then in my scientific diary, a notebook I use to scribble down anything that might be useful for a later article or book.

At the time, I was staying at a chalet in the Ardennes, owned by a couple of friends. In the early morning hours, as the rising sun illuminated the surrounding woods, I opened my diary to write down thoughts that had spun during the night. Perhaps it was the peace and quiet of the natural environment that made me more sensitive than usual, but on that November morning, I was gripped by the palpable and acute awareness of a new totalitarianism that had left its seed and made the fabric of society stiffen.

Even by 2017, it could no longer be denied: The grip of governments on private life was growing tremendously fast. We were experiencing an erosion of the right to privacy (especially since 9/11), alternative voices were increasingly censored and suppressed (particularly in the context of the climate debate), the number of intrusive actions by security forces was rising dramatically, and more.

It was not only governments behind these developments, however. The rapid emergence of "woke" culture and the growing climate movement was giving rise to the call for a new, hyper-strict government that emerged *from within the population itself.* Terrorists, climate changes, heterosexual men, and, later, viruses were considered too dangerous to be tackled with old-fashioned means. The technological "tracking and tracing" of populations became increasingly acceptable and was even deemed necessary.

The dystopian vision of the German-Jewish philosopher Hannah Arendt loomed at society's horizon: the emergence of a new totalitarianism, no longer led by flamboyant "mob leaders" such as Joseph Stalin or Adolf Hitler but by dull bureaucrats and technocrats.

That November morning, I drafted the blueprint for a book in which I would explore the psychological roots of totalitarianism. At the time, I wondered: Why did totalitarianism as a form of statehood emerge for the first time in the first half of the twentieth century? And moreover: What is the difference between it and the classical dictatorships of the past? The essence of this difference, I realized, lies within the field of psychology.

Dictatorships are based on a primitive psychological mechanism, namely on the creation of a climate of fear amongst the population, based on the brutal potential of the dictatorial regime. Totalitarianism, on the other hand, has its roots in the insidious psychological process of *mass formation*. Only a thorough analysis of this process enables us to understand the shocking behaviors of a "totalitarized" population, including an exaggerated willingness of individuals to sacrifice their own personal interests out of solidarity with the collective (i.e., the masses), a profound intolerance of dissident voices, and pronounced susceptibility to pseudo-scientific indoctrination and propaganda.

Mass formation is, in essence, a kind of group hypnosis that destroys individuals' ethical self-awareness and robs them of their ability to think critically. This process is insidious in nature; populations fall prey to it unsuspectingly. To put it in the words of Yuval Noah Harari: Most people wouldn't even notice the shift toward a totalitarian regime. We associate totalitarianism mainly with labor, concentration, and extermination camps, but those are merely the final, bewildering stage of a long process.

* * *

In the months and years after I made these initial notes, more and more references to totalitarianism appeared in my diary. They spun into longer and longer threads that organically connected with other areas of my academic interests. For example, the psychological problem of totalitarianism touched upon a crisis that had erupted in the scientific world in 2005, a theme that I explored extensively in my doctoral dissertation. Sloppiness, errors, biased conclusions, and even outright fraud had become so prevalent in scientific research that a staggeringly high percentage of research papers—up to 85 percent in some fields—reached radically wrong conclusions. And the most fascinating thing of all, from a psychological point of view: Most researchers were utterly convinced they were conducting their research more or less correctly. Somehow, they failed to realize that their research was not bringing them closer to the facts but instead was creating a fictitious new reality.

This, of course, is a serious problem, especially for contemporary societies that place their faith in science as the most reliable way of understanding the world. Furthermore, the foregoing problem is directly related to the phenomenon of totalitarianism. In fact, this is precisely what Arendt exposes: The undercurrent of totalitarianism consists of blind belief in a kind of statistical-numerical "scientific fiction" that shows "radical contempt for facts": "The ideal subject of totalitarian rule is not the convinced Nazi or the convinced Communist, but people for whom the distinction between fact and fiction and the distinction between true and false no longer exist."[1]

The poor quality of scientific research reveals a more fundamental problem: Our scientific worldview has substantial shortcomings, the consequences of which extend far beyond the field of academic research. These shortcomings are also the origin of a profound collective unease, which, in recent decades, has become increasingly palpable in our society. People's view of the future is now tainted with pessimism and lack of perspective, more so everyday. Should civilization not be washed away by rising sea levels, then it certainly will be swept away by refugees. The Grand Narrative of society—the story of the Enlightenment—no longer leads to the optimism and positivism of yesteryear, to put it mildly. Much of the population is trapped in almost complete social isolation; we see a remarkable increase in absenteeism due to mental suffering; an unprecedented proliferation in the use of psychotropic drugs; a burnout epidemic that paralyzes entire companies and government institutions.

In 2019, this predicament was clearly perceptible in my own professional environment. I saw so many colleagues around me drop out from work due to psychological problems, hindering the capacity to perform even basic day-to-day work. For example, that year, it took me nearly nine months to obtain a signature on a contract that was required for me to get started on a research project. The university departments that had to review the contract and grant their approval were dealing with so much absenteeism that there was always someone on sick leave due to mental suffering, so that the contract simply didn't get finalized. During that period, all social stress indicators rose exponentially. Anyone familiar with systems theory knows what this means: The system is heading for a tipping point. It is on the verge of reorganizing itself and seeking a new equilibrium.

At the end of December 2019—in the same Ardennes chalet I mentioned earlier—I ventured to make a small prediction in the company of friends: One of these days, we will wake up in a different society. This intuitive premonition even enticed me to take action. A few days later, I went to the bank to pay off the mortgage on my house. Whether or not that was a wise thing to do depends entirely on your perspective. Maybe it wasn't wise from a purely economic or tax point of view, but that was of no concern to me. First and foremost, I wanted my sovereignty back; I did not want to feel indebted to and complicit in a financial system that, in my view, played a part in the social impasse that was about to occur. The bank manager listened to my story; he even agreed with me. But he insisted on knowing why I felt so determined about it. Even after we spoke for an hour and a half, it wasn't enough to fill the emptiness of his question. I ended up leaving him wondering, well past the closing time of his branch office, which was to be shut down forever shortly thereafter.

* * *

A few months later—in February 2020—the global village began to shake on its foundations. The world was presented with a foreboding crisis, the consequences of which were incalculable. In a matter of weeks, everyone was gripped by the story of a virus—a story that was undoubtedly based on facts. But on which ones? We caught a first glimpse of "the facts" via footage from China. A virus forced the Chinese government to take the most draconian measures. Entire cities were quarantined, new hospitals were built hastily, and individuals in white suits disinfected public spaces. Here and there, rumors emerged that the totalitarian Chinese government was overreacting and that the new virus was no worse than the flu. Opposite opinions were also floating around: that it must be much worse than it looked, because otherwise no government would take such radical measures. At that point, everything still felt far removed from our shores and we assumed that the story did not allow us to gauge the full extent of the facts.

Until the moment that the virus arrived in Europe. We now began recording infections and deaths for ourselves. We saw images of overcrowded emergency rooms in Italy, convoys of army vehicles transporting corpses, morgues full of coffins. The renowned scientists at Imperial College confidently predicted that without the most drastic measures, the virus would claim tens of millions of lives. In Bergamo, sirens blared day and night, silencing any voice in public space that dared to doubt the facts. From then on, story and facts seemed to merge and uncertainty gave way to certainty.

The unimaginable became reality: We witnessed the abrupt pivot of nearly every country on Earth to follow China's example and place huge populations of people under de facto house arrest, a situation for which the term "lockdown" was devised. A surreal silence descended—ominous and liberating at the same time. The sky without airplanes, traffic arteries without rushing blood; the dust of chasing vain desires settling down, and in India, the air became so pure that, for the first time in thirty years, in some places the Himalayas became once more visible against the horizon.[2]

It didn't stop there. We also saw a remarkable transfer of power. Expert virologists were called upon as George Orwell's pigs—the smartest animals on the farm—to replace the unreliable politicians. They would run the animal farm with accurate ("scientific") information in a time of plague. But these experts soon turned out to have quite a few common, human flaws. In their statistics and graphs, they made mistakes that even "ordinary" people would not easily make. It went so far that, at one point, they counted *all* deaths as coronavirus deaths, including people who had died of, say, heart attacks.

Nor did they live up to their promises. These experts pledged that the Gates to Freedom would reopen after two doses of the vaccine, but when the time came, things didn't change and they came up with the need for a third. And just like Orwell's pigs, they sometimes changed the rules overnight, inconspicuously. First, the animals had to comply with the measures because the number of sick people could not exceed the capacity of the health care system (*flatten the curve*). But one day, everyone woke up to discover writing on the walls stating that the measures were being extended because the virus had to be eradicated (*crush the curve*). Eventually, the rules changed so often that only the pigs seemed to know them. And even that was not so sure.

Some people became suspicious. How is it possible that these experts make mistakes that even laymen wouldn't make? Aren't they scientists, the kind of people who took us to the moon and gave us the internet? They can't be that stupid, can they? What is the endgame? Their recommendations take us further down the road in the same direction: With each new step, we lose more of our freedoms, until we reach a final destination where human beings are reduced to QR codes in a large technocratic medical experiment.

That's how most people eventually became certain. Very certain. Yet of the most opposing things. Some people were convinced that we were dealing with a killer virus, others that it was nothing more than the seasonal flu, and still others believed that the virus did not even exist and that we were dealing with a worldwide conspiracy. And there were also a few who continued to tolerate uncertainty and kept asking themselves: How can we adequately understand what is going on in our society?

* * *

The coronavirus crisis did not come out of the blue. It fits into a series of increasingly desperate and self-destructive societal responses to objects of fear: terrorists, global warming, coronavirus. Whenever a new object of fear arises in society, there is only one response and one defense in our current way of thinking: increased control. The fact that the human being can tolerate only a certain amount of control is completely overlooked. Coercive control leads to fear and fear leads to more coercive control. Just like that, society falls victim to a vicious circle that inevitably leads to totalitarianism, which means to extreme government control, eventually resulting in the radical destruction of both the psychological and physical integrity of human beings.

We have to consider the current fear and psychological discomfort to be a problem in itself, a problem that cannot be reduced to a virus or any other "object of threat." Our fear originates on a completely different level—that of the failure of the Grand Narrative of our society. This is the narrative of mechanistic science, in which man is reduced to a biological organism. A narrative that ignores the psychological, symbolic, and ethical dimensions of human beings and thereby has a devastating effect at the level of human relationships. Something in this narrative causes man to become isolated from his fellow man, and from nature; something in it causes man to stop *resonating* with the world around him; something in it turns the human being into an *atomized subject*. It is precisely this atomized subject that, according to Arendt, is the elementary building block of the totalitarian state.

Totalitarianism is not a historical coincidence. In the final analysis, it is the logical consequence of mechanistic thinking and the delusional belief in the omnipotence of human rationality. As such, totalitarianism is the defining feature of the Enlightenment tradition. Several authors have postulated this, but it hasn't yet been subject to a psychological analysis. This book fills that gap. We will analyze the symptom of totalitarianism and situate it within the broader context of the social phenomena of which it forms a part.

Part 1 (chapters 1 to 5) covers how the mechanist–materialist view of man and the world creates the specific social-psychological conditions in which mass formation and totalitarianism thrive. Part 2 (chapters 6 to 8) details the process of mass formation and its relationship to totalitarianism. Finally, part 3 (chapters 9 to 11) investigates a way to transcend the current condition of man and the world, so as to render totalitarianism superfluous. As a matter of fact, part 1 and part 3 of this book only marginally refer to totalitarianism. It is not my aim with this book to focus on that which is usually associated with totalitarianism—concentration camps, indoctrination, propaganda—but rather the broader cultural-historical processes from which totalitarianism emerges. This approach allows us to focus on what matters most: Totalitarianism arises from evolutions and tendencies that take place in our day-to-day lives.

Ultimately, this book explores the possibilities of finding a way out of the current cultural impasse in which we appear to be stuck. The escalating social crises of the early twenty-first century are the manifestation of an underlying psychological and ideological upheaval—a shift of the tectonic plates on which a worldview rests. We are experiencing the moment in which an old ideology rears up in power, one last time, before collapsing. Each attempt to remediate the current social problems, whatever they may be, on the basis of the old ideology will only make things worse. One cannot solve a problem using the same mindset that created it. The solution to our fear and uncertainty does not lie in the increase of (technological) control. The real task facing us as individuals and as a society is to construct a new view of man and the world, to find a new foundation for our identity, to formulate new principles for living together with others, and to reappraise a timely human capacity—speaking the truth.

PART I

SCIENCE AND ITS PSYCHOLOGICAL EFFECTS

CHAPTER 1

Science and Ideology

It is a summer day in 1582. A young student by the name of Galileo Galilei sits in the cathedral of Pisa. Before him stands a priest, reciting scripture. A chandelier attached to the vaulted ceiling by a thin chain hangs over the priest's head. The warm summer breeze blows in through the open doors, setting the chandelier in motion. Sometimes, the breeze swings the lamp far from its resting place above the altar; other times it moves it only a little bit. The priest's voice disappears into the background. Galileo's eyes follow the lamp—back and forth, back and forth. He checks his pulse and counts the number of heartbeats. Regardless of how far it swings, the pendulum always takes the same amount of time to return to its starting point.

The events in the cathedral of Pisa later took on mythical proportions, embodying the cultural and social upheaval that characterized the centuries that followed. Religious discourse, with its system of dogmas derived from ancient texts, lost its authority. Instead of something that had to be revealed to man by God, knowledge became something man could come to on his own. All he had to do was observe phenomena with his eyes and think logically.

Religious discourse had turned man's gaze inward for thousands of years, revolving around the conception of man as a sinner, who lies and deceives and loses himself in worldly temptations, who must ready himself for death because it will catch up with him eventually. If man suffered in this world, the creation of God, it was because he failed to measure up as a moral and ethical being, because he was living in sin. It was not the world that had to be questioned but man himself.

That all changed with the emergence of science: Man believed that, with the power of reason, he could adjust the world, while he himself could remain unchanged. He gathered his courage and took charge of his destiny: He would use his own intellectual power to understand the world and to shape a new, rational society. For too long, he had been forced to remain silent in the name of a God no one had ever seen; for too long, society had been burdened by dogmas that lacked any rational foundation. The time had come to dispel the darkness with the light of reason. "Enlightenment is man's release from his self-incurred tutelage. Tutelage is man's inability to make use of his understanding without direction from another . . . 'Dare to think! Have the courage to use your own reason!' is therefore the motto of the Enlightenment," as stated in 1784 by the great German Enlightenment philosopher Immanuel Kant.[1]

Galileo dared—to think. After Mass, he rushed to his dorm room and began experimenting with pendulums: He altered the weight of the swinging object, the force with which the object was put into motion, the length of chain by which the object was suspended. Only a few months later, he was able to formulate the basic law governing pendulum motion: Only the length the chain (the pendulum arm) has an impact on the duration of motion.

Other brilliant thinkers, such as Nicolaus Copernicus and Isaac Newton, also pulled the dogmatic wool from their eyes to register the world around them with an open mind. They demonstrated that certain aspects of reality could be captured in mathematical and mechanistic formulas with incredible accuracy and precision. It seemed incontrovertible: The book of the universe is written in the language of mathematics.

These thinkers not only reached great intellectual achievements, they also assumed a unique humanistic and ethical stance with regard to the world and its material objects. They had the courage to set aside the prejudices and dogmas of the time. They admitted their ignorance and were curious and open to what phenomena have to say for themselves. This "not knowing" gave birth to a new knowledge, a new knowledge for which they would do anything, for which they were willing to give up their freedom, sometimes even their lives.

This newborn science—this budding knowledge—showed all the characteristics of what the French philosopher Michel Foucault defines as *truth-telling*.[2] Truth-telling is a way of speaking that breaks through an established, if implicit, social consensus. Whoever speaks the truth breaks open the solidified story in which the group seeks refuge, ease, and security. This makes speaking the truth a dangerous endeavor. It strikes fear in the group, and results in anger and aggression.

Truth-telling is dangerous. Yet it is also necessary. No matter how fruitful a social consensus may be at a certain time, if it is not dismantled in time and renewed, it will putrefy and eventually have a suffocating impact on society. In such times, the truth will emerge as a sincere voice that breaks through the dull refrain of an established story and lends a new sound to old and ageless words. "Le vraie est toujours neuf" (Truth is always new) (Max Jacob).[3]

Science can, in essence, be defined as open-mindedness. The original practice of science, that which formed the basis of the Enlightenment, briefly suspended prejudice about the things being observed. It was open to the greatest possible diversity of ideas and thoughts, assumptions, and hypotheses. It cultivated doubt and considered uncertainty a virtue. It let the facts speak for themselves and decide for themselves what kind of thought or theory they preferred to unite with. In this way, facts were reborn into words as fresh, budding truths.

It was not only the facts that had the liberty to assert themselves. "I may disagree with what you say, but I will defend to the death your right to say it," Voltaire declared (or rather, his biographer Evelyn Beatrice Hall, declared). Science also liberated man from his self-incurred immaturity. It broke through rule by religious dogma that, in the public sphere, had largely decayed into coercion and oppression, pretense and hypocrisy, deceit and lies.

This open-mindedness bore abundant fruit. The scientific method was used to understand and predict the movement of the celestial bodies, to describe pendulums and calculate gravitational acceleration, and also to study the behavior of animals, to understand how the mind works, to map the structure of languages, to compare cultures with one another. It could be flexibly adapted to every domain of inquiry, every object of research, and it brought forth sublime discoveries in every field. Shapes and colors were delineated sharper than ever in science's light; sounds sounded clearer than ears had ever heard.

This openness of mind, this faithful pursuit of Reason at any cost eventually yielded, through incessant endeavor over several centuries, the most sublime insights. *Surprising* insights, too. The great physicists of the first half of the twentieth century proved in the most rigorous way that the core of matter cannot be separated from the observing subject. They demonstrated that the observation of a material object changes the object itself ("Looking at something, changes it," Erwin Schrödinger declared).

Moreover, they relinquished the illusion that man could ever attain certainty. With his uncertainty principle, Werner Heisenberg demonstrated that it is impossible to unambiguously determine even purely material "facts," such as the location in time and space of material particles.[4] The great minds who followed reason and facts most rigorously came to the conclusion that, ultimately, the essence of things is beyond logic and cannot be grasped. Niels Bohr concluded that only poetry can describe the absurd behavior of elementary particles: "When it comes to atoms, language can only be used as poetry."

Also the very idea of predictability in the material world—fanatically proclaimed by French scientist Pierre-Simon Laplace in the eighteenth century—was invalidated by the American mathematician and meteorologist Edward Lorenz in the twentieth century. Even if you're able to capture a complex and dynamic phenomenon (which includes most natural phenomena) in mathematical formulas, you still, even with formulas in hand, wouldn't be able to predict their behavior a second in advance.

And finally, the image of the universe as a dead and nondirectional (nonteleological) mechanical process also proved scientifically untenable.

Chaos theory showed in a truly revolutionary way that matter is constantly organizing itself in ways that cannot possibly be explained in mechanistic terms. The universe is endowed with direction and *volition*. We'll explore this more in detail in the last part of this book.

Newton had already stated as much in the seventeenth century: The laws of mechanics apply to only a very limited part of reality. As science progressed, this only became clearer—at least, for those who had eyes to see it. In the twentieth century, the great mathematician René Thom put it this way: "That portion of reality, which can be well described by laws which permit calculations, is extremely limited." He continued, even more importantly, "All major theoretical advances, in my opinion, have arisen from the capacity of their inventors to 'get into the skin of things,' to be able to empathize with all entities of the external world. It is this kind of identification that transforms an objective phenomenon into a concrete thought experiment."[5]

This sheds surprising light on the nature of science. Most are of the opinion that science consists of making dry, logical connections between "objectively" observable facts. However, science is, in fact, characterized by *empathy*, a resonant affinity between the observer and the phenomenon under investigation. As such, science stumbles upon an unknowable and mysterious essence that escapes logical explanation and which can be described only in the language of poetry and metaphor.

Encounters with that essence often result in what we might describe as the seminal religious experience—a religious experience that precedes and is untainted by religious institutions or dogma. Max Planck testified to that experience, in perhaps the most direct and vulnerable way: Science eventually arrives where religion once started, in a personal contact with the Unnameable (see also chapter 11).

Based on this experience, the physicists of the twentieth century reappraised the great religious and mystical writings, such as the *Upanishads*. The content and structure of those texts, the imagery and the symbolism, offer a better grip on reality than any logical, rational discourse. Science freed itself from all the dogmas of religious discourse, only to rediscover—at the end of a long journey—the mystical and religious texts and reendow them with their resplendent, original status: symbolic, metaphorical texts for that which is eternally hidden from the human mind.

As we will discuss in the latter part of this book, the faithful pursuit of Reason attained the highest and most sublime achievement: mapping its own boundaries. The human mind had accepted its own limitations and once more relocated the ultimate knowledge beyond and outside itself. The ultimate achievement of science is that it finally surrenders, that it comes to the realization that it cannot be the guiding principle for man. It is not human reason that is at the heart of the matter, but man as an individual who makes ethical and moral choices, man in relation to fellow man, man in relation to the unnameable, which, at the heart of things, speaks to him.

* * *

However, from the beginning, the tree of science also sprouted a branch in another direction—the exact opposite direction of that original scientific practice. Based on the great achievements of science, some people tipped from open-mindedness to belief; for them, science became ideology. It was mainly the mechanistic-materialistic branch, the so-called hard sciences, that most enraptured us. Simple in its principles (the laws of mechanics), specific in its object (the tangible, visible world), and awe-inspiring in its practical applicability (from the steam engine to television and the atomic bomb to the Internet), this science has everything to seduce human beings. With this branch of science, man conquers space; it enables us to see and hear what is happening on the other side of the planet and visualize brain activity; it gives us the ability to move faster than sound and to perform microsurgical procedures. In the past, people waited in vain for God to perform miracles, but this science made them actually happen. Man had left the stage of believing and could now rely successfully on what he *knew*. At least, so he *believed*.

From the Enlightenment forward, mechanistic thinking provided the Grand Narrative in Western civilization. According to that story, it begins with a big bang that sets an expanding universe in motion, generating a series of phenomena of growing complexity. Hydrogen is formed first, then helium, and then all the other elements through alternating processes of fusion and explosion. The elements clump together and form stars and planets and one of them, the Earth, contains water. This water allows for the formation of amino acids, often regarded as the first form of life. From here, guided by natural selection, simple forms of life gradually give way to more complex forms until, at long last, man emerges—the provisional end point of evolution. In this way, the scientific discourse spun its own creation myth.

From this perspective, the entirety of human subjectivity becomes an insignificant by-product of mechanistic processes.

> *Man may not realize it, but his humanity does not really matter, it is nothing essential. His whole existence, his longing and his lust, his romantic lamentations and his most superficial needs, his joy and his sorrow, his doubt and his choices, his anger and unreasonableness, his pleasure and his suffering, his deepest aversion and his most lofty aesthetic appreciations, in short, the entire drama of his existence, can ultimately be reduced to elementary particles that interact according to the laws of mechanics.*

This is the credo of mechanistic materialism.

"Whoever doubts this creed, voluntarily declares himself foolish or insane." One is still allowed to doubt, but only about the "right" things. In this way, the tree of science sprouted a branch that grew in the opposite direction from the original shoots. At its birth, science was synonymous with open-mindedness, with a way of thinking that banished dogmas and questioned beliefs. As it evolved, however, it also turned itself into ideology, belief, and prejudice.

Science thus underwent a transformation, as all ideologies do. At first, it was a discourse by which a minority defied a majority; then it became the discourse of the majority itself. In the course of this transformation, scientific discourse aligned itself with objectives that were opposed to the original ones. It enabled manipulation of the masses, allowed people to build a career ("publish or perish"), promote products ("Research shows our soap washes the whitest"), spread deception ("I only believe the statistics I faked myself," Winston Churchill), and belittle and stigmatize others ("Whoever believes in alternative medicine is an irrational fool"). Indeed, even to justify segregation and exclusion (no access to public spaces unless you bear the sign—a mask, a vaccine passport—of the scientific ideology). In short, the scientific discourse, like any dominant discourse, has become the privileged instrument of opportunism, lies, deception, manipulation, and power.

* * *

To the extent that the scientific discourse became an ideology, it lost its virtue of truth-telling. Nothing illustrates this better than the so-called replication crisis that erupted in academia in 2005. This crisis emerged when a number of serious cases of scientific fraud came to light. Scientific scans and other imaging were proven to have been manipulated,[6] archaeological artefacts were found to be counterfeit,[7] embryo clones had been forged;[8] some researchers claimed to have successfully transplanted skin from mice, whilst they had simply dyed the skin of the test animals without performing any surgical procedure.[9] Other researchers had manufactured missing links from pieces of skulls of humans and monkeys;[10] and yes, it appeared that some even completely made up their research.[11]

This kind of full-fledged fraud was relatively rare, however, and not actually the biggest problem. The biggest problem was with less dramatic instances of questionable research practices, which were reaching epidemic proportions. Daniele Fanelli conducted a systematic survey in 2009 and found that at least 72 percent of researchers were willing to somehow distort their research results.[12] On top of that, research was also replete with unintentional calculation mistakes and other errors. An article in *Nature* rightly called it "a tragedy of errors."[13]

All of this translated into a problem of replicability of scientific findings. To put it simply, this means that the results of scientific experiments were not stable. When several researchers performed the same experiment, they came to different findings. For example, in economics research, replication failed about 50 percent of the time,[14] in cancer research about 60 percent of the time,[15] and in biomedical research no less than 85 percent of the time.[16] The quality of research was so atrocious that the world-renowned statistician John Ioannidis published an article bluntly entitled "Why Most Published Research Findings Are False."[17] Ironically, the studies that assessed the quality of research also came to diverging conclusions. This is perhaps the best evidence of how fundamental the problem is.

In recent decades, academics have attempted to improve the quality of research through a number of initiatives. They questioned the pressure on researchers to publish, urged researchers to make their data publicly available, pushed for more transparency around financial interests, and more. Overall, these measures don't seem to have had much effect. In 2021, 50 percent of surveyed academics anonymously admitted that they sometimes presented their findings in a biased way. Half is already a problem, but according to Fanelli, it almost certainly represents a substantial underestimation. This is because a significant percentage of the researchers, even if surveyed anonymously, will not admit to engaging in questionable research practices. The measures taken to improve the quality of scientific research, however well-intentioned, failed to address the problem.

The replication crisis does not simply indicate a lack of seriousness and scrupulousness in research. It first and foremost points to a fundamental epistemological crisis—a crisis of the way in which science is conducted. Our interpretation of objectivity is wrong, excessively based on the idea that numbers are the preferred approach to facts. If we look at the scientific fields with the worst replicability outcomes, it becomes clear that the *measurability* of phenomena plays a significant role. In chemistry and physics, for example, it wasn't that bad. However, in psychology and medicine, the situation is wretched. In those fields, researchers assess extremely complex and dynamic phenomena—the physical and psychological functioning of human beings. Such "objects" are, in essence, only measurable to a very limited extent, as they cannot be reduced to unidimensional characteristics (see chapter 4). And yet, all too often, we see desperate attempts to mold them into data.

In both medicine and psychology, measurement is usually done on the basis of tests that result in numerical scores. These figures give the impression of being objective; however, this needs some perspective. Studies into so-called "cross-method agreement" start from a question that is as simple as it is interesting: If you measure the same "object" using different measurement methods, to what extent will the results coincide? If the measurement methods are accurate, the results should be virtually identical. However, this is not the case. Not even close. In psychology, for example, the correlation between the results obtained by different measurement methods rarely exceed 0.45. This, of course, is an abstract number, which is why I like to give a concrete example in my university lectures. Imagine you are building a house and a carpenter comes to take measurements for eight windows. He uses three different tools on each window: a folding rule, a tape measure, and a laser measure. If the carpenter's measurements are as inadequate as a psychologist's, he would report the following results (see table 1.1).

Table 1.1. A Carpenter's Measurements with a Psychologist's Accuracy

	Folding Rule (in cm)	**Tape Measure (in cm)**	**Laser Measure (in cm)**
Window 1	180	130	60
Window 2	100	200	150
Window 3	160	220	130
Window 4	100	170	210
Window 5	30	100	20
Window 6	120	80	160
Window 7	110	150	60
Window 8	30	90	10

With the folding rule, the carpenter concludes that window 1 is 180 cm wide; with the tape measure, the same window is 130 cm wide; and with the laser measure, it is 60 cm wide. It is the same scenario with the second window: The folding rule shows that window 2 is 100 cm wide, the tape measure shows that it is 200 cm wide, and the laser measure shows that it is 150 cm wide. The correlation among all sets of the three measurements is 0.45.

Would you hire this carpenter? This is about the best you can expect when psychologists use three different measuring instruments. This doesn't mean that all psychological measurements are meaningless, but the idea that they are "objective" needs to be put into perspective.[18]

As a young researcher, I intended to tackle the measurement problem, thinking that only the field of psychology was burdened with this problem to such an extent. I later discovered that it applies equally to medical sciences (and many other fields of science, as well, as we'll see in chapter 4). The tests and measuring instruments in medicine are—this may surprise you—on average no better than those used in psychology. Take a look at the in-depth survey study by Gregory Meyer and his colleagues.[19]

During the coronavirus crisis, the public became aware—perhaps for the first time—of the relativity of medical measurements, as we witnessed the manifest problems with the PCR test. It quickly became clear that the test can be administered in different ways, that it produces widely variable results, that the results can also be interpreted in different ways, and so on. Johann Goethe once said, "Measuring a thing is a crude act, which cannot be applied in any other way than extremely imperfectly to living bodies." By attempting to measure the unmeasurable, measurement becomes a form of pseudo-objectivity. Instead of bringing the researcher closer to his research object, the measurement procedure leads him further away. It hides the examined object behind a screen of numbers.

Low-validity tests and data collection methods are not only problematic per se; they also prevent a researcher from attempting to understand his object in a different, maybe less sophisticated-looking, but often more appropriate way, just by using means of words, for example. This is the real drama of fields like medicine and psychology: They have abandoned the classic research, such as thorough case studies conducted by experienced clinicians, and replaced it with research that might look scientific but often is not. Metrical data might seem like a more sophisticated and objective way of describing the research object, but it often conveys less than a skillful description by means of words. This led, in part, to the other problems that surfaced in the scientific crisis: the ubiquitous errors, sloppiness, and biased conclusions, which we talked about earlier. Anyone who tries to squeeze the unmeasurable into numbers will sense that his research has little real value and will be less motivated and lack a sense of duty to deliver accurate work.

The lack of quality in scientific research raises a few pressing questions, including about the blind peer review system, which is used in all scientific journals and is considered the ultimate seal of approval for scientific legitimacy. Peer review requires that a study be read and critically evaluated by two or three independent experts in the field before publication. These experts are supposed to be "blind" (they don't know who conducted the study), but in reality, they usually do know the authors because they know the other researchers working in their field. Hence, they can usually guess who conducted the research. For this reason, a fair assessment by an expert requires not only that he is willing and able to free up sufficient time and energy—far from given in the current academic climate. Moreover, it requires that he is capable of identifying his personal prejudices with regard to the research and its authors, and put them aside. In other words: Peer review stands or falls on the ethical and moral quality of the expert—that is, his subjective, human characteristics.

* * *

And just like that, this chapter has come full circle. Both great Science (the science that maintains an open mind and pursues Reason) and small science (the science that degenerates into ideology) eventually re-encounter what they originally had pushed out of view: man as a subjective and ethical being. The first kind of science does so in a positive way, by recognizing the importance of that dimension and anchoring it in its theories. It started as a courageous, young science by looking outwards to the material world, registering phenomena and establishing logical connections between them. It assumed—and rightly, to a degree—that this was the way to sovereign knowledge. In great Science, the human being, in its psychic, symbolic, moral, and ethical dimension, disappeared into the background. But that didn't last long. It was discovered that the observer, in his subjective qualities, has an essential influence on the objects being observed. The theories in which those insights have been anchored, such as quantum mechanics and complex, dynamic systems theory, have to be considered among the greatest achievements man has ever produced. (We will explore this in more detail in part 3.)

To the extent that science has degenerated into ideology, belief, and dogma—small science—it has also confirmed that the human being, in its subjective dimension, is the central point of focus. In this case, however, science does so in a negative way, by testifying to this with its own failure. It increasingly ignored the register of subjective experience, eventually considering it to be a kind of insignificant, quasi-unreal by-product of material, biochemical processes in the brain, for example. But that didn't make the subjective dimension cease to exist. It proliferated, took on grotesque proportions, and manifested itself as a torrent of errors, sloppiness, questionable research practices, and outright fraud. Ultimately, human subjectivity also reclaimed its throne in small science as well.

As we will discuss more extensively in chapter 3, the most striking thing of all is that, in general, researchers themselves hardly realize that there is something wrong with their methodology. They generally take their scientific fiction for reality, confusing their numbers with the facts of which they are a distorted echo. The same applies to a large part of the population, blindly trusting this scientific ideology, with no other ideological hiding place, given the fall of religion. Numbers and graphs presented in the mass media by someone with credentials are considered de facto realities by many people. It is at this level that Hannah

Arendt situates the ideal subject of the totalitarian state: the subject that no longer knows the difference between (pseudo)-scientific fiction and reality. Never before were there so many such people as in the beginning of the twenty-first century; never before were the societal conditions so prone to totalitarianism.

CHAPTER 2

Science and Its Practical Applications

Science not only leads to knowledge and intellectual advances, it also has effects in the real world through its practical applications. Mechanistic science, in particular, had high ambitions in this regard. It wants to adapt the world to people, to make life easy and comfortable, and ultimately eliminate suffering and even death.

To a certain extent, science also fulfilled those ambitions. Galileo's discovery allowed Christiaan Huyghens fifteen years later to build a mechanical device to measure time: the pendulum clock. Until then, people mainly depended on natural cycles to measure time; now, people were able to create artificial cycles of any duration by altering the length of the pendulum arm. As such, a day could be broken down into 86,400 identical pendulum seconds. Time changed from an elusive stream of natural cycles into a quantifiable process, hopping forward in strictly identical mechanical steps.

What came next was an almost endless series of practical applications: the steam engine, the camera, artificial light, the radio, the television, the automobile, the airplane, the internet. In the two centuries following Newton's formulation of the basic laws of motion—no more than a blink of an eye in human history—society became mechanized and industrialized in a dizzying number of ways. During thousands of years, man had been subjected to the world; now he imposed his will upon it. For the first time, he was able to radically change his troubled condition and make life easier. Or at least, he had that impression.

Yet there was, undeniably, another side to the coin. Each added convenience came at a price, including a weakened connection to the natural and social environment. Artificial light broke the rhythm that the sun and moon had hitherto imposed on daily activities; the clock separated the human mind from cyclical natural processes (meeting up as soon as the dew has dried, eating when the sun is at its highest point, going to sleep when the night falls); the compass alienated man from the stars; industrial labor drew him away from the fields and the woods. The psychological impact of all this usually wasn't considered important—if it was even considered at all. But it was undoubtedly immense. Prior to mechanization, man's world of experience constantly resonated with nature's ever-varying language of forms; after mechanization, he was mainly absorbed by a monotonous, mechanical rhythm.

Social connections were also transformed beyond recognition. The invention of radio and television led to the rise of the mass media and a corresponding decline in direct human interactions with a merely social function. Evening meetings between neighbors, pub gatherings, harvest festivals, rituals, and celebrations—they were progressively replaced by consumption of what the media presented. This seduced us into certain social laziness. It was no longer necessary to make the effort that is required for interaction with fellow human beings.

No risk of arguing; no confrontation with painful jealousy, shame, or embarrassment; no need to dress up or to even leave the house. It also *uniformized* social exchanges. Public space, including the political sphere, was increasingly dominated by a shrinking number of voices that conquered the living room via the mass media.[1] In other words, social relationships lost their diversity and originality.

The mechanization of the labor process also engendered a profound transformation of social structures and connections, a dimension explored by Marx's historical materialism. The steam engine, for example, could power such a large number of looms and provide employment to such a large number of people that new forms of society, such as factory villages, rose around it. These communities were merely focused on mass production, wage labor being the only point of collective identification. As such, industrialization broke up traditional social structures formed by the existence of varied professions, public offices, and authority (the priest, the mayor). Although these structures curbed man's freedom for centuries, or even radically suppressed it, they also offered him a psychological basis and frame of reference. They gave him rules and laws, commandments and prohibitions, boundaries to his lusts and urges, well-defined objects of anxiety, frustration, and anger. Their disappearance left man confused, in the darkness of his own existence; haunted by existential anxiety and unease that could not be identified. As we will see in chapter 6, this unfettered anxiety plays a crucial role in mass formation and totalitarianism.

The mechanization of the world also had a direct effect at the level of meaning making. Mass production rendered the end result of labor less tangible. In the past, man worked to produce the objects needed to sustain the bodily existence of oneself and the people around him. He worked to feed himself, to warm the house, to clothe himself against harsh conditions and the gaze of others. That changed with the rise of the industrial environment. He now worked to produce objects—for people far away. The answer to the question of what is the meaning of one's work no longer welled up from one's own body.

In addition, the Other for whom one worked was anonymous. The effect of one's work on the Other could no longer be seen or felt. With the disappearance of (much of the) local, small-scale, and craft production, the direct link between producer and consumer was broken. In most cases, the person who produced the material good no longer came into contact with the person who was about to use it. When a product was delivered, the person who produced it no longer witnessed the joy or gratitude on the recipient's face. It's these visible, subtle physical effects that primarily provide human satisfaction in work; they are the most direct sign that work is meaningful. In this way, not only one's own body but also the other faded as sources of meaning making. The worker became, as they say, a cog in the industrial machine, lubricated only by the thought of wages due. Labor changed from a cumbersome but inherently meaningful existential task into a disembodied utilitarian necessity.

* * *

Besides waning meaning, another problem arose. Surprisingly, industrialization and mechanization of labor didn't mean that less work needed to be done. In the early twentieth century, British economist John Maynard Keynes predicted that by the end of the century, technological advances would translate to a 15-hour work week, which would be sufficient for society to produce everything it needed.[2] He was correct on that last point—more than correct, in fact. It probably requires even fewer than fifteen hours of labor for society to achieve that. But his prediction didn't come true. By the end of the twentieth century, people worked longer hours than ever before.

What Keynes failed to consider was the creation of meaningless and useless work on an incredible scale. Professor of anthropology David Graeber described this in his by now well-known book *Bullshit Jobs*. He asked a random sample of people whether they thought their jobs made a meaningful contribution to society. About 37 percent answered with a definite "no" and an additional 13 percent were unsure.[3] These bullshit jobs were mostly created in the administrative and economic sectors, and the countless occupations that support these sectors. Graeber tells the story of "Kurt," who works at a company providing auxiliary services for the German army, and illustrates the degree of absurdity that gradually began to characterize so many people's working lives, and existence:

> ***Kurt:*** *The German army hires a subcontractor for their IT work. The IT company hires a subcontractor who takes care of the logistics side. The logistics company hires a subcontractor for their personnel management, and I work for that company.*

Suppose a soldier moves to an office two doors down the hallway. Instead of simply picking his computer up and taking it there, he has to fill out a form.

The IT company receives the form, people read it and approve the application, and send it to the logistics company. The logistics company then approves the computer to be moved to the office two doors down the hallway and asks us for staff. My company's office workers then do their thing, and that's where I come in.

I receive an e-mail: "Come to barrack C at time B." Usually those barracks are about a hundred to three hundred miles from my house, so I rent a car. I drive the rental car to the barracks, I let the dispatcher know that I have arrived, fill out a form, disconnect the computer, put the computer in a box, seal the box, ask someone from logistics to carry the box to the room five meters further down the hallway, I reopen the box there, fill out another form, reconnect the computer, call the coordinator to let him know how long it took me, have a few people sign off, drive my rental car home, send all the paperwork to the coordinator and get paid.

So instead of the soldier being allowed to move his computer five meters further down the hallway, two people have to drive a total of six to ten hours, fill out about fifteen forms and waste more than four hundred euros in tax money.[4]

This is an intriguing aspect of the phenomenon of meaningless work: You'd think that in private companies, dominated by capitalist pursuits and dictated by profit, such absurd work wouldn't exist. Why would a for-profit company hemorrhage money on unprofitable workers? However, this idea can be relegated to the realm of illusions.[5] Even in the private sector, there is a proliferation of meaningless work. We can attribute this in the first place to the changes in corporate culture. Today's executives rarely have a true personal stake in the success or failure of the company they lead. They can afford to create pointless jobs, perhaps to do friends a favor, or to give the company a sophisticated image by employing any manner of "experts," if need be even solely to optimize their employment statistics. By the time the company goes bust, the executive will have been employed elsewhere for a while anyway.

But there's more to it than that. The rampant growth of the administrative and economic sectors has to do with much more fundamental psychological tendencies in our society. Endless proliferation of rules, procedures, and administration usually stems from interpersonal mistrust and inability to tolerate uncertainty and risk. Both the government and the population are ever more demanding that everything be done *correctly*. This involves endless procedural provisions, necessary to determine who is financially and legally liable if anything goes wrong. As we will discuss in chapter 5, today's compulsion to regulate and control is a frenetic attempt to master ever-growing anxiety.

If human relationships are characterized by fundamental distrust, life becomes hopelessly complicated and society spends its energy at creating all kinds of "security mechanisms," which in fact fuel mistrust even more and are, above all, psychologically exhausting. That's why the phenomenon of bullshit jobs is also directly associated with the epidemic of workplace burnout. What makes work performance unbearable, is usually not the actual demands but the impossibility of experiencing meaning and satisfaction, of experiencing work as an act of *creation*. Put someone in an office and pay him a generous wage to perform a useless task, like pushing a button every ten minutes. Does such a job free you from the burdens of life, or does it make your life unbearably light?

In the end, a paradox arises: feelings of resentment and revenge toward those who have meaningful work. It's remarkable that it's mainly the people who perform work that is directly useful—health care workers, garbage collectors, craftsmen, farmers—who get fired or whose work is so poorly rewarded that they have to almost live on the breadline or survive from subsidies (think of farmers, who produce food, the most necessary material object of all). On the other hand, the most meaningless jobs, such as administrative work, are steadily increasing in number and are, in comparison, rewarded more and more generously. This is more or less the (unconscious) reasoning of: "If you are lucky enough to have a meaningful job, you should not expect to be adequately rewarded on top of that." And just like that, we've ended up in a situation in which it almost seems foolish to choose meaningful work.

The rise of meaningless professions shows us that the real problem of humanity lies in human relationships, more so than in the struggle with natural forces or in the physical demands of work. Simply put, in a society in which human relationships are satisfying, life will be bearable even if it has only primitive means of production. Whereas in a society where human relationships are impoverished and toxic, life will be difficult and unbearable, however "advanced" such society may be in terms of mechanical-technological evolutions.

* * *

To summarize, science led to a formidable ability to alter the material world through industrialization and mechanization. But this also gave rise to problems, especially regarding our relationships, both with each other, and with nature. Furthermore, we're faced with problems that are caused by the fact that science—or that which passes for science today—is often neither accurate nor reliable.

In chapter 1, I explained that the quality of research is most problematic in medical science. No less than 85 percent of medical studies come to questionable conclusions due to errors, sloppiness, and fraud. This allows us to understand, for example, why drugs that are found to be safe in research trials may, in practice, cause thousands of deaths, or generate significant side effects. The most well-known example might be the thalidomide scandal. Thalidomide (Softenon) was marketed in 1958 as an anti-nausea medication for pregnant women. By 1961, it was clear that thalidomide had caused severe malformations in at least ten thousand fetuses, mostly underdeveloped limbs or the absence of limbs altogether. The most mind-boggling aspect of the scandal is that pharmaceutical companies continued to produce the drug for years, and that in some countries (including Belgium), it was sold over the counter until 1963. This drug that deformed thousands of babies and destroyed thousands of lives wasn't withdrawn from the market until 1969. The justification is perplexing, to put it mildly: The government first wanted to be 100 percent sure that there was, indeed, a link between the drug and fetal malformations.

Another dramatic example concerns the artificial hormone diethylstilbestrol (DES), which was widely administered between 1947 and 1976 to prevent miscarriages. Around 1976, it became clear that the use of DES was a terrible mistake. It did not prevent miscarriages, but it did have a series of serious side effects that affected multiple generations.[6] The women who took it developed a higher risk of breast cancer. The first generation of female offspring were at higher risk of abnormalities in the endometrium, pregnancy complications, genital deformations, and an increased risk of cervical, breast, and vaginal cancer. The first-generation of male offspring were at increased risk of nodules on the epididymis, while the second-generation of male offspring had a higher rate of ureteral abnormalities. Nobody knows if, and in which generation, the abnormalities caused by DES will cease to exist.

Thalidomide and DES are probably the most well-known medical scandals, but they're not the ones that resulted in the greatest number of victims. In 2019, a massive lawsuit was filed against several pharmaceutical companies for their role in the opioid crisis, killing as many as four hundred thousand people over the past twenty years and ruining untold millions of American lives. One takeaway from this tragedy is that even pharmaceutical drugs that enjoy long-term and widespread use aren't necessarily safe. Only in 2021, it was discovered that the popular painkiller acetaminophen (Tylenol), which has been on the market since 1955, contains carcinogens and can be harmful to fetuses.

But aren't the effects and side effects of pharmaceutical drugs tested extensively before they are brought to market? How is it possible that all these harmful side effects are not discovered? Here is the problem: The phenomenon of "health" or "reaction to a drug" is a complex and dynamic phenomenon that cannot possibly be measured or understood in its entirety. A researcher can only record and monitor a very limited number of responses (for example, the effect on the symptom, the effect on blood pressure, or respiration). He remains largely in the dark about everything else. Additionally, research is only conducted for a limited period of time. The side effects that manifest after that period, even generations later, such as with thalidomide, can't be fully accounted for. And finally, side effects can also be too subtle to detect immediately but quite serious over time, such as a decrease in general immunity.

Accurate prediction is further complicated by strong psychological factors. The placebo effect (where a treatment has positive effects, merely because the patient believes in its effectiveness) and nocebo effect (where a treatment has negative effects because the patient believes it is harmful) are widely accepted phenomena. And they're not minor, as some might say. Some researchers (such as Shapiro[7] and Wampold[8]) estimate that up to 90 percent of the effects of medical treatments can be attributed to psychological factors. If this is correct, most medical treatments would more accurately be described as (unacknowledged) psychotherapy.

Although these data, like all data, are relative, it is clear that the influence of psychological factors is significant (chapter 10 is completely dedicated to this). That's why the effects of pharmaceuticals and medical interventions are difficult to predict, and they can also change over time as the zeitgeist changes. Different discourse leads to different expectations and different expectations lead to different effects. This helps explain why drugs appear to lose their initial efficacy after being on the market for a while. A new therapy often raises high expectations, creating a strong placebo effect. Only from a naïve mechanistical perspective does one believe that the effects of medical interventions can be objectively measured through experiments.

The poor quality of medical research also raises pressing ethical questions. For instance, it shines a harsh light on the merciless drive to conduct experiments. Every year the number of laboratory animals used for medical experiments increases.[9] In 2005, about one hundred million animals were sacrificed worldwide (!); by 2020, this nearly doubled to just under two hundred million (!). The fate of these animals is horrific, often too horrific for words. If we take into account that 85 percent of medical studies are erroneous, biased, or even fraudulent (see chapter 1), we can only conclude that, in the majority of cases, this inferno of suffering is meaningless and unnecessary on top of that. Where exactly do we draw the line between experimentation and torture? If such a practice reaches such magnitude and such a degree of absurdity in a society, we cannot but conclude that such a society is seriously ill.

* * *

Mechanistic thinking gave man an enormous capacity to manipulate the material world. Combined with the (self-) destructive tendency intrinsic to man, this has put him in the most precarious situation he has ever been in. For the first time in history, man is able to raze the "natural resources" on which he depends, depleting the world's fish stocks, for example, and clearing entire rainforests. Furthermore, with the industrialization and mechanization of war, mechanistic thinking showed its destructive potential in an overt and direct way. The tens of millions of victims of the destruction machines that were deployed in the world wars are silent witnesses thereof. And even more so in the years to follow, the sinister marriage between science and murderous rage wreaked such havoc that the war misery of yesteryear paled in comparison. To give just one example, Monsanto produced seventy-six million liters of Agent Orange, which was sprayed in Vietnam to defoliate the trees and drive the Vietcong out of the jungle. The result? Millions of both Vietnamese and American soldiers became seriously ill, often with tumors and cancers, causing deformities in at least 150,000 children.

While mechanistic science sought to make the human condition more comfortable, in many respects it also made it more dangerous. Man could not help but feel threatened by the powers he himself unleashed from nature. And, for the most part, those powers ended up in the hands of a few. Due to the industrialization, mechanization, and technologization of the world, production capacities, economic power (via a self-centralizing banking system), and psychological power (via mass media) fell into the hands of an ever-decreasing number of people. The Enlightenment tradition had promised people autonomy and freedom, but, in a way, it brought people greater (feelings of) dependence and powerlessness than ever before. This powerlessness caused people to increasingly mistrust those in power. Throughout the nineteenth century, fewer and fewer people felt that political leaders really represented their voice in public space or defended their interests. As a result, man also became disassociated from the social classes that were represented by the politicians and was left uprooted, no longer connected to the whole of society, no longer belonging to a meaningful social group.

Although the Enlightenment tradition arose from man's optimistic and energetic aspiration to understand and control the world, it has led to the opposite in several respects: namely, the experience of loss of control. Humans have found themselves in a state of solitude, cut off from nature, and existing apart from social structures and connections, feeling powerless due to a deep sense of meaninglessness, living under clouds that are pregnant with an inconceivable, destructive potential, all while psychologically and materially depending on the happy few, whom he does not trust and with whom he cannot identify. It is this individual that Hannah Arendt named the *atomized subject*. It is this atomized subject in which we recognize the elementary component of the totalitarian state.

CHAPTER 3

The Artificial Society

What is the endgame of the mechanistic ideology? To answer this question, we must return to the cathedral of Pisa where the eyes of seventeen-year-old Galileo Galilei follow a swinging lamp. With his youthful openness and curiosity, Galileo sees something that countless eyes had never noticed: Whether the pendulum makes a long or a short swing, the time it takes to swing hence and forth is always the same. Upon closer analysis, this makes sense. Long swings start from a higher position and as the object begins its downward motion, it accelerates in its path. Shorter swings start from a lower position, and as the object begins its downward motion, it accelerates less. The speed at which the pendulum travels on its path is directly proportional to the length of the arc it makes—and therefore the movement of the pendulum always lasts the same amount of time.

Galileo's discovery was brilliant, no doubt. But it wasn't quite right. Christiaan Huyghens noticed something when he was building his pendulum clocks: If he attached several clocks to the same wall, their pendulums would eventually move in a perfect simultaneous manner.[1] He couldn't help but conclude that somehow the clocks were in communication with one another. Huyghens assumed—rightly, as it turned out—that the vibrations of the pendulums spread through the wall, causing small deviations in duration that, in a way that is difficult to understand, eventually cause the pendulum movements synchronize.

That is to say, pendulums are more complex than Galileo's simple law suggests. Apparently, they have the ability to adjust their movements under the influence of their environment. Precision measurements of the duration of motion confirm Huyghens's view, at least to the following extent: Contrary to what Galileo thought, pendulums do not always swing for exactly the same amount of time. Sometimes it takes just a little longer, sometimes just a little less time to complete its movement.[2] And this also turned out to be the case if a pendulum is swinging in an isolated state, without the process of synchronization: The swings' durations are not exactly the same. Initially, these deviations were dismissed as a form of insignificant "noise." The irregularity in the pendulum was believed to be the result of coincidental mechanical factors, such as changes in surrounding airflow or the chain twisting.

It took until the second half of the twentieth century to discover that this is not correct. These apparently random deviations form a pattern that can be described with a mathematical formula but is nevertheless strictly unpredictable. (Pendulums have the characteristic of deterministic unpredictability, which we will revisit in chapter 9). What's more, the aforementioned pattern is unique to each pendulum. Pendulums had been regarded as dull, mechanical phenomena that dutifully followed Galileo's laws, but those elementary mechanical devices were, in fact, creative in nature and idiosyncratically capable of disobedience. In *Chaos*, James Gleick puts it this way: "Those studying chaotic dynamics discovered that the disorderly behavior of simple systems acted as a creative process. It generated complexity: richly organized patterns, sometimes stable and sometimes unstable, sometimes finite and sometimes infinite, but always with the fascination of living things."[3]

Reducing the pendulum's behavior to Galileo's law robs it of its "social" qualities, as well as its individuality and creativity. If you were to create a virtual pendulum in a computer program that behaves strictly according to Galileo's law, it would look very much like a real pendulum, but it would be a death phenomenon, lacking the lively chaos of a real pendulum.

* * *

Galileo's pendulum illustrates a universal law: The logic and rational explanation of a natural phenomenon—however comprehensive it may be—always makes an abstraction of that phenomenon. Theoretical models never capture anything fully; they always leave an unexplained remainder. This remainder is not just insignificant, random "noise." It is the essence of the object. It is its living component.

You can see this, for example, in the difference between "natural" and "artificial" products. Whether it's a genetically engineered plant, lab-printed meat, vaccine-induced immunity, or high-tech sex dolls—whenever we artificially reproduce a natural phenomenon from rational analysis, the artificial phenomenon is not identical to the original. The loss is not always immediately visible. Sometimes it is barely visible at all. And yet, it is crucial, both on a physical and psychological level. The digitalization of human interactions—replacing real human interactions with digital ones—is a good example thereof.[4]

With the coronavirus crisis, the trend toward a digital society made a big leap forward. Teleworking became the norm, student life took place online,[5] aperitif and coffee were consumed in front of a television or computer screen,[6] even sex was mediated through technological machinery[7] and the death penalty was carried out from a safe digital distance.[8] Initially, it was mainly seen as a necessity and occasionally as an advantage. People felt protected from the virus, saved time, avoided traffic jams, reduced their ecological footprint, and spared themselves the stress and discomfort that can characterize human encounters.

However, this acceleration of online existence also accelerated burnout and exhaustion, to the extent that some now speak of *digital depression*.[9] Perhaps the heart of the problem lies in the following: A conversation not only conveys information; there is also a subtle but equally profound bodily exchange and this is disrupted by digitalization. This physical aspect of speaking is of vital importance. It makes language a matter of love and lust, charged with a refined eroticizing power. That's why we physically crave a real conversation after a week of working online.

A digital conversation is not the same as a real conversation. We see this most clearly in infants. During the first six months, they learn to distinguish language sounds at an astonishing pace, but only while listening to someone who is physically present, not when listening to an audio or video recording (see Kuhl's experiments[10]). Early language learning is inseparable from the physical presence of the "other." The child internalizes the mother's (body) language, as it satisfies its physical needs with the warmth of her body, the milk of her breasts. The child breathlessly fixates on the mother's face and imitates the expressions that play on it; it listens with the closest attention to the sounds she makes and even with its earliest sobbing and crying already echoes the melody and tones of her speech.

What's more: This synchronization already takes place before birth, in the womb. Annie Murphy Paul's experiments ("What babies learn before they are born"[11]) show that the infant's crying immediately after birth already bears melodic resemblance to its mother's voice. And if a newborn listens to its mother's voice through headphones while nursing on the left breast and someone else's voice while nursing on the right, it will begin nursing significantly more on the left. The conclusion is inescapable: The child has already become familiar with its mother's voice in the womb; life in the womb has predestined it to resonate with that specific voice.

After birth, the child further develops this primal resonance. This doesn't happen haphazardly. The child achieves a kind of *symbiosis* with the mother through its creative imitations of her sounds and facial expressions; in this way, it will feel what she feels. As it takes on its mother's happy expression, it also feels her joy; if it takes on her sad expression, it shares in her unhappiness. Something similar applies to the exchange of sounds: In the clinking and clanging of the mother's language trembles the well and woe of her being, and the child who imitates that language resonates with it on the same psychological wavelength.

This early resonance between child and its (social) environment leads to a unique phenomenon: The young child's body gets "loaded" with a series of vibrations and tensions that become embedded in the deepest and finest fibers of its body. They form a kind of "body memory" that not only programs the function of the musculature, glands, nerves, and organs, but also predisposes the child to certain psychological conditions, or disorders.

The human body is, in the most literal sense, a *stringed instrument*. The muscles that span the skeleton, and the body's other fibers, are put on a certain tension in early childhood through imitative language exchanges. This tension determines with which (social) phenomena one will resonate; it determines the frequencies to which one will be sensitive in later life. That's why certain people and certain events can literally strike a chord; they touch the body and, as such, touch the soul. It is for this reason that the voice can make the body ill. Or, conversely, heal it.

That is why the voice is of vital importance, especially at an early age. Lack of a voice is fatal to the young child. The Austrian-American psychiatrist René Spitz studied two groups of children whose biological needs (food, drink, clothing, housing) were satisfied in identical ways, except that one group had a stable psychological bond with a caretaker and the other did not. Spitz found that the mortality rate was significantly higher in the latter group.

This subtle physical dimension of linguistic exchange remains important throughout life. While speaking, adults, like young children, constantly mirror the facial expressions and postures of their interlocutor without even realizing it (see research into the so-called mirror neurons).[12] This happens through a kind of inner imitation, through slight and imperceptible increases in muscle tensions. No matter how subtle, this is more than enough to gauge, in an immeasurably short time span, the deeper layers of the other's subjective experiencing—whether that person is in pain, feels sad or happy, is perhaps just pretending—and to mimic it.

This leads to a remarkably direct connection between interlocutors. Professionally, I have been studying (psychotherapeutic) conversations in detail for fifteen years and have been able to ascertain this in a concrete way. To highlight a single aspect: People react incredibly quickly to one another during conversations. When one person stops speaking, the other usually begins in less than 0.2 seconds (the response time to a traffic light is, on average, five times longer). And this happens even if the speaker doesn't finish his sentence, so that the other person cannot possibly predict when he will stop on the basis of the semantic structure of the sentence.

When people talk to one another, they sense each other very sharply because they perceive the slightest changes in intonation, voice timbre, facial expression, body position, rate of speech, and so on. Like flocking starlings, they form one organism. They are connected with one another through a psychic membrane that transfers the slightest ripple in body and soul. In every exchange of words, no matter how trivial, people show themselves to be perfect dance partners; they are subtly united through the eternal music of language. We make love more often than we realize.

This complex phenomenon degrades when digitized. Digital interactions always have a certain delay; exclude certain aspects of contact, such as scent and temperature; are selective (you see only someone's face); and create the constant, unpleasant preapprehension that the connection may drop. As a result, digital interactions are not only experienced as reticent and stiff; they also give us the feeling that we cannot really (physically) sense the other. In the words of workplace leadership expert Gianpiero Petriglieri: "In digital interactions, our minds are tricked into believing that we are together, but our bodies know that we are not; what's so exhausting about digital conversations is being constantly in the presence of the other person's absence."[13]

From here, we see a direct association between digitalization and depression. In classical psychoanalytic theory, depression is associated with the frustrating experience of helplessness, induced by the passivity or absence of a loved one (usually a parent, in childhood).[14] Subsequently, you pay the "Other" with the same currency: You yourself become passive (i.e., depressed). Digital "connection" leads to a similar dynamic: You feel helpless with respect to an Other, whom you experience as absent and unreachable, and react with frustration and passivity (i.e., feeling exhausted).

Digitalization *dehumanizes* a conversation. This usually happens in a hidden, insidious way, but sometimes it can also be felt very sharply. A recent example from my psychotherapeutic practice: A woman in her early forties wakes up one night with her hands covered in blood and realizes she is miscarrying the baby she's been yearning for her entire life. She asks me, sobbing, for a conversation—a *real* conversation. In such a situation, anyone can sense that the digital wall will not be scalable for the words in which the drama seeks its expression. Unless there really isn't any other possibility, offering a digital conversation in such a situation seems indeed almost inhumane.

Similar examples can be extracted from educational settings (the enthusiasm of the teacher, which is almost physically palpable in a classroom does not tolerate the journey through a fiber optic cable); work environments (the support of a project leader is diluted in an online meeting); love life (try to salvage a wavering love, with all the linguistic torment that characterizes it, through online communication); and actually any situation to which a person must be fully accompanied by his humanity.

If all this is true, then why are digital interactions so *attractive*? Why did we happily give up chitchatting for text messages, long before the coronavirus crisis? It is convenient to communicate in this way with people who are far away; this is certainly true. However, there is also another, psychological factor at play. Uncertainty is the preeminent characteristic of human experience—no other animal is so haunted by doubt or plagued by existential questions—and this is especially true of our relationship to the Other. How can I do good for the Other? Does he like me? Does he find me attractive? Do I mean something to him? What does he want from me?

In a digital conversation, in which the Other is literally kept at a distance but can still be reached, these eternal questions and the associated uncertainty and fear become less acute. The sense of control is far greater; it's easier to selectively show some things and hide others. In short, people feel psychologically safer and more comfortable behind a digital wall but pay a price for it with the loss of connectedness. This brings us to a theme that will recur repeatedly in this book: The mechanization of the world causes man to lose contact with his environment and become an atomized subject, the kind of subject in which Hannah Arendt recognized the essential component of the totalitarian state.

* * *

Science adapts its theory to reality, whereas ideology adapts reality to theory. This includes mechanistic ideology, which attempts to adapt reality to its theoretical fiction. It aims to optimize nature and the world. We already mentioned genetically engineered plants and animals, lab-printed meat, and other artificial products, but it extends much further than that. Some argue that menstrual periods are a superfluous inconvenience and advocate for eliminating them with artificial hormones and turning the female cycle into a single, flat line.[15] And after years of experimenting with "growing" cow and dog fetuses in an artificial womb,[16] which is little more than a plastic bag (see figure 3.1), some people believe that it's also time to replace a mother's womb with a synthetic sack.[17]

The only thing missing to make such practices completely identical to the breeding programs in Aldous Huxley's *Brave New World* is for the mother's voice to be replaced by the monotonous repetition of conditioning messages. In such a case, the melodious echoes of the mother's voice will no longer be reflected in her newborn's cries. Instead, the baby will arrive in the world, already "socially adapted." Other advantages cannot be underestimated. The future parents will be able to continue their normal lives during the nine months of "pregnancy."[18] It isn't yet entirely clear as to whether the presence of the child will be allowed to change life at all after the synthetic womb opens and the child is "born."

The synthetic womb is not as far away as we think. The only thing required to persuade a society that is gripped by the mechanistic ideology is a slew of "experts" daily presenting statistics and data in the media, informing us that artificial wombs protect fetuses a few percentage points better against viruses and pathogens than the not-so-sterile mother's body. Within this logic, anyone who chooses natural pregnancy will be considered unfit as a parent—such people would expose their child to unnecessary risks, even before birth. Whether

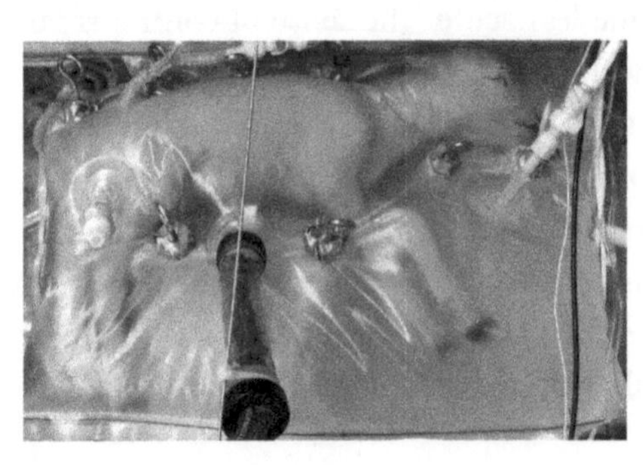

Figure 3.1.

dissident voices could override such logic remains to be seen. Life itself can be defended only in terms of metaphor and poetry, yet these usually sound less loud than the monotonous droning of mechanistic arguments.

These trends fit within the broader vision of an ideal society. Institutions that preoccupy themselves with the society of the future, such as the World Economic Forum, consider it a matter of course that the world will move toward a digicosm, a "society" in which human life is mainly conducted online. Strangely enough, the twenty-first century environmental movement is following this trend in lockstep. Insofar as it travels the "ecomodernist" route, it aims to save nature by protecting it from man. In those terms, living in the countryside is a crime, just like lighting a wood stove and eating a piece of real (read: not lab-printed) meat. Within such logic, the ideal life is spent indoors, on an intravenous drip. The fact that man and nature form a mystical unity and can exist in harmony is considered to be a romantic and unrealistic idea, even downright dangerous, considering the pressing issue of the climate change.

This social vision tends to intersect with so-called transhumanism. This is a contemporary iteration of the mechanistic ideology that considers it desirable, even necessary, that future humans merge physically and mentally with machines. Transhumanists want to replace the chaos of writhing bodies with a strictly technological *internet of bodies*. To this end, bodies have to be saturated with microchips and be monitored via a powerful internet. Once this is achieved, it will not only be possible to fight crime and sexual harassment more efficiently than ever before, but also to carry out genetic correction and preventive medicine through the collection of biometric data and replacing the body's natural resilience with vaccine generated artificial immunity. Even the human mind would benefit from these developments. In 2020, Elon Musk announced that, within five years, we will no longer need clumsy human language—that source of eternal misunderstanding—because he will provide a microchip that can be built into the brain and that will allow humans to communicate via flawless digital signals.[19]

What follows should come as no surprise: Within this utopia, they also want to control weather conditions—that source of angst for farmers worldwide since time immemorial—by means of radical mechanical-technological means. Such measures are considered indispensable due to global warming, and the technologists believe they can do so. For example, they can obscure the sun by placing "smart" mirrors between the Earth and the sun, by launching sulphate clouds from rockets, or by detonating chalk bombs in the stratosphere.[20] Mechanistic ideology always lives on credit! In the future, once perfect knowledge has been achieved and perfect technology has been mastered, it will translocate the man-machine into paradise. Yet for now, it mainly makes people sick and depressed.

The triumphant music of the mechanistic ideology always contains a discordant note. If we know anything by now, it's that the achieved convenience always comes at a price, and that price usually becomes apparent only after it's too late. The fluorine compounds in Teflon pans and the PFAS in water repellent raincoats turn out to be carcinogenic.[21] So is the ethylene oxide used in hundreds of everyday products.[22] The connection between chemicals and chronic, noninfectious, degenerative diseases, the so-called diseases of civilization, is basically well known, but that doesn't stop or redirect the relentless drive to push "civilizing" further down that road.[23] The greater the impact of mechanistic science on the world, the more it becomes clear that we're creating problems for which we can hardly find a solution. The ever-thickening plastic soup in the oceans and the nuclear waste that remains active for hundreds of thousands of years are just a few examples. Those problems were, in principle, clear from the start to those who had the eyes to see it. In the eighteenth century, the British painter and poet William Blake, for example, already had a keen sense of the destructive and dehumanizing nature of the mechanization of the world. In a sense, his entire oeuvre testifies to it. Unfortunately, he was, and remains, an exception.

Why is mankind so hopelessly seduced by the mechanistic ideology? Partly because it's under the influence of the following illusion: that one is able to remove the discomforts of existence without having to question oneself at all. This is best illustrated by modern medicine. The cause of suffering is typically traced to a mechanical "defect" in the body or to an external entity, such as a pathogenic bacterium or virus. Its cause is localized and can (in principle) be controlled, managed, and manipulated without the patient having to wrestle with any psychological, ethical, or moral complexity. "A pill helps you to get rid of your problems," "Plastic surgery frees you from your complexes without having to question the origin of your shame and embarrassment." While the practical applications of mechanistic science make life easier, in a sense, the essence of life eludes us ever more. Much of that process takes place below conscious awareness, but the sharp increase in acute mental suffering is an unmistakable sign that is discernible at society's surface.

The Enlightenment man could hardly help but cling to utopian optimism. In the nineteenth century, industrialization heralded the disappearance of the aristocratic and class society, and associated local social structures. Man tumbled out of his social and natural context, and as he fell, meaning dropped away too (see chapter 2). In this "disenchanted" mechanistic world (Max Weber), life became meaningless and a-teleological (the machinery of the universe runs without meaning or purpose), and religious frames of reference also lost coherence.[24] Anxiety and unease, once tied to the oppression and abuse of the aristocracy and clergy, began to drift ineffably around in the human soul. Frustration and aggression, once held in check by fear of hell and the last judgment, proved increasingly easy to mobilize. The prospect of an afterlife dwindled and was readily replaced by belief in an artificially created, mechanistic-scientific paradise.[25]

It is here that we, together with Hannah Arendt, situate the undercurrent of totalitarianism: a naive belief that a flawless, humanoid being and a utopian society can be produced from scientific knowledge.[26] The Nazi idea of creating a purebred superman based on eugenics and social Darwinism, and the Stalinist ideal of a proletarian society based on historical-materialism are prototypical examples, as is the current rise of transhumanism. When we hear about such ideologies, we like to believe that they are the products of deranged minds. This is a misconception. Plato, for example, found eugenics a commendable practice that had a place in his ideal state.[27] And the twentieth century taught us that this practice does indeed lead to certain "successes." The systematic abortion of fetuses with genetic predispositions to thalassemia in Cyprus resulted in this hereditary blood disease almost completely disappearing from the island.

We have to seriously ask ourselves the following question: Why *not* follow the principles of eugenics? As a social strategy, it can be rejected on purely ethical grounds, but it is crucial that we also be capable of rejecting it on rational grounds. The essence on rational grounds might be this: Eugenics may sometimes lead "locally" to desired results, insofar as it concerns "combating" "undesirable" characteristics; from an overall point of view, however, it has more disadvantages than advantages. Government regulation of the intimate sphere leads to psychological despair and, ultimately, to a decline in physical health. (We will further elaborate on this theme in the final chapters.) Even within the context of an ideology that would make physical health its ultimate goal, eugenics is a questionable strategy that ignores the complexity and subtlety of the human being.

As Hannah Arendt states, totalitarianism is ultimately the logical extension of a generalized obsession with science, the belief in an artificially created paradise: "Science [has become] an idol that will magically cure the evils of existence and transform the nature of man."[28] In the next chapter, we will delve more deeply into one of the core features of both the mechanistic and totalitarian discourse: a naive belief in the measurability of reality and the excessive use and misuse of data and statistics.

CHAPTER 4

The (Im)measurable Universe

In the chapter 3, we subjected the (utopian) *goal* of mechanistic ideology to critical analysis. In this chapter, we will focus on the *method* this ideology uses to gather knowledge. The universe is a machine, the components of which are measurable—that is the basic assumption of this ideology. Measurements and calculations form the basis of the mechanistic research methods. This epistemological point of departure has bearing on the ideology's conception of the ideal society. Ideally, society is led by expert technocrats who make decisions based on objective, numerical data. With the coronavirus crisis, this utopian goal seemed very close at hand. For this reason, the coronavirus crisis is a case study par excellence in subjecting the trust in measurements and numbers to critical analysis.

Until this recent crisis, societies were not primarily governed on the basis of numerical data. They were guided by *stories*, first by mythical and religious stories and later by political stories. The mechanistic ideology cannot accept this trust in stories because they are essentially irrational and subjective in nature; they say more about the author of the story than about any so-called objective reality it represents. Stories consist of words, words that can mean anything; they have no solid, rational relationship to facts.

And without a rational basis, man drifts astray—or so mechanistic ideology believes. Ultimately, all these stories usually favor their authors; think of the indulgences of the clergy and the no-show jobs granted to politicians. We should not take this lightly. It leads to the abuse of power, or eventually to absurd horror. The ritually burned widows of India and the drowned witches of Europe are but a few silent witnesses from an endless array of victims. This is the way in which past societies went from bad to worse: stories – subjectivity – irrationality – poignant injustice – absurd horror.

The coronavirus crisis offered an unexpected window of opportunity for the mechanistic ideology: The uncertainty and fear of the virus provided a basis for the formation and development of a society in which decisions are based on numbers rather than stories. Today, we are talking about relatively "simple" numbers on infections, hospitalizations, and deaths; in the future, we might be talking about high-tech, biometric data that precisely map every aspect of physical function.

Unlike words, numbers offer an objective basis for transparent and rational decisions. As such, they are an antidote to the abuse of power and absurd horror. Moreover, they offer an opportunity to minimize human suffering. This is the path to the rational society of the future: data – objectivity – rationality – accuracy – minimization of suffering. In this light, the coronavirus could become the crowning achievement of humanity. At least, that's more or less how the story goes.

Have a look at figure 4.1. If you measure the length of the coastline of Great Britain based on a unit of measurement of 200 kilometers, it is 2,400 kilometers long. If you measure it with a unit of 50 kilometers, it is 3,400 kilometers long. As you decrease the unit of measurement, the length of the coastline of Great Britain increases to infinity. The reason is simple: As the measurement unit becomes smaller, it more closely follows the irregular coastline and the border becomes

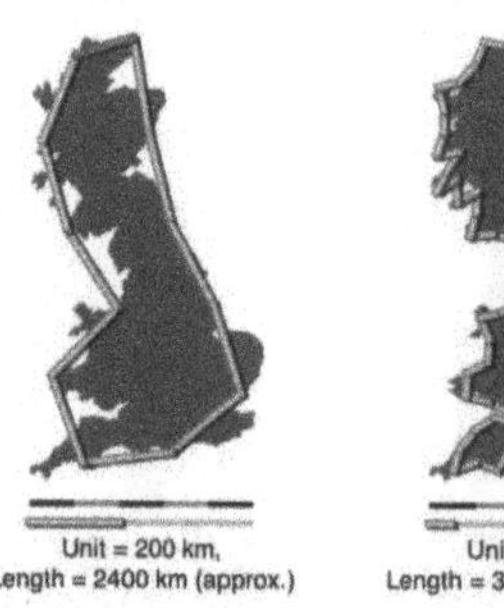
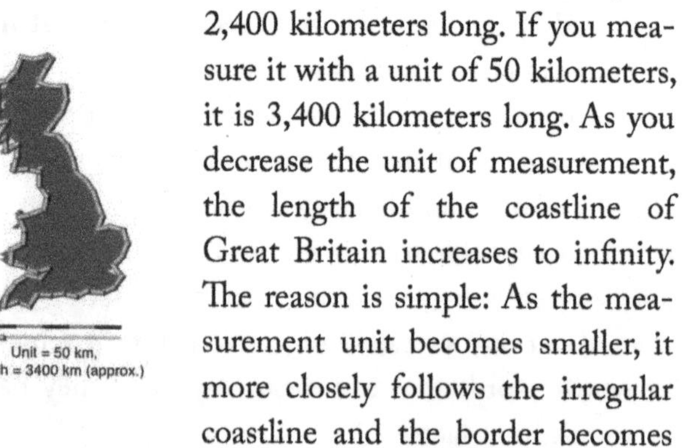

Figure 4.1.

longer. This is how the brilliant Polish-Jewish mathematician Benoit Mandelbrot showed that measurements are always relative, depending on a series of subjective choices, such as the unit of measurement.[1]

* * *

And even in the rare cases where measurements themselves can be considered accurate and quasi-objective (for example, measuring the length of strictly unidimensional objects, such as a stick, or counting members of discrete categories), there still is an important subjective factor at the level of interpretation. This is illustrated by an example known in statistics as Simpson's paradox.[2] Table 4.1 shows the number of executions carried out for the crime of murder in the state of Florida, divided between white and black offenders. The conclusion is clear: White people are more likely to receive the death penalty than black people in

Table 4.1. Executions in Florida by Race of the Offender

Race offender	Capital punishment		Execution percentage
	YES	NO	
WHITE	19	141	11.9
BLACK	17	149	10.2

Table 4.2. Executions in Florida by Race of the Victim

Race offender	Race victim	Capital punishment		Percentage
		YES	NO	
WHITE	WHITE	19	132	12.6
	BLACK	0	9	0
BLACK	WHITE	11	52	17.5
	BLACK	6	97	5.8

Florida. Researchers concluded that the prejudice against black people must be incorrectly attributed as a driver of capital punishment—that is, until a statistician presented the same numbers in a slightly different way. He not only divided the perpetrators' race into white and black; he also divided the victims' race accordingly (see table 4.2). This led to the opposite conclusion.

Black people are more likely to get the death penalty if they kill a white person than white people are if they kill a black person. It's tempting to think that this is the final analysis, but there is no doubt that the numbers can be presented in a still different way, which could lead to still different conclusions.

Numbers have a unique psychological effect. They create an almost irresistible illusion of objectivity, which is further enhanced when numbers are presented visually in charts or graphs. When people see numbers, they believe them to be *objects* or *facts*. This illusion blinds people to the nonetheless obvious truth that numbers are always relative and ambiguous, that they are constructed and produced from an ideologically—and subjectively—shaded story. At first glance, the numbers seem only true to the facts, yet on closer inspection, it becomes clear that they slavishly serve every story.

* * *

In chapter 1, we saw that the so-called replication crisis that erupted in the sciences in 2005 was never really resolved. Since then and up until now, science has continued to struggle with an epidemic of errors, sloppiness, forced conclusions, and fraud. The coronavirus crisis was, in a sense, just a continuation of this crisis. The difference is that this time the spectacle did not take place within academia but openly in the public square. All the problems that had surfaced a decade prior now played out in the mass media, in plain sight, in front of the world. Many people could hardly believe their eyes and ears when they witnessed scientists at the highest levels contradicting themselves and their colleagues, making simple calculation and counting errors, changing their minds injudiciously, being transparently influenced by financial interests in their scientific pronouncements, even openly admitting that they had deliberately misled the population.

Numbers played a crucial role in this saga. In principle, the coronavirus crisis was about calculating relatively simple phenomena, such as the number of infections, hospitalizations, and deaths. However, it was plain as day that the data were anything but objective. The number of infections was usually determined by PCR tests, which didn't run smoothly. These tests are designed to determine whether RNA sequences from a virus are present in the body.[3] Those RNA sequences can come from a virulent virus but also from a "dead" virus. As a result, people may, even months after an infection (and therefore long after they are contagious), still test positive. And this was just one of the many limitations of the test.

Estimating the change in infection rate based on the positive test result rate also proved very problematic. Public health experts who spoke to the media about infection trends, for example, stubbornly refused to adjust for the total number of tests performed. (In technical terms, they reported the absolute number of positive tests instead of the positivity ratio.) In the summer of 2020, virologist and former rector of the University of Liège, Bernard Rentier, was given access to raw data about the so-called summer wave (called, at that time, the second wave). He subjected these data to a critical analysis and concluded that the estimated number of infections after adjustment for the total number of tests performed was between *twenty to seventy times lower* than the estimates reported in the media.[4] If you think mistakes like this could only be made once, you would be wrong. In the summer of 2021, the scenario repeated itself. This time, the positivity ratio was occasionally mentioned, but once again, we were warned of a summer wave based on graphs depicting the absolute number of infections.

Data on hospital admissions were also extremely relative. Throughout the crisis, any patient who tested positive upon admission was considered a COVID-19 patient, regardless of whether they had COVID-19 symptoms or, let's say, a broken leg. At a certain point, the Scottish government changed its methodology and began counting someone as a coronavirus patient only if they tested positive and were also admitted with COVID-19 symptoms. The result? They were left with *13 percent* of the original number of COVID-19 patients.[5]

This was not the only factor that distorted hospital data. In the spring of 2021, Jeroen Bossaert of the Flemish newspaper *Het Laatste Nieuws* published one of the few thorough pieces of investigative journalism of the entire coronavirus crisis. Bossaert exposed that hospitals and other healthcare institutions had artificially increased the number of deaths and COVID-19 hospitalizations for financial gain.[6] This in itself is not surprising, since hospitals have been using such methods for a long time. What *was* surprising is that, during the coronavirus crisis, people refused to acknowledge that profit motives played a role and had an impact on the data. The entire healthcare sector was suddenly graced with quasi-sanctity. This, despite the fact that prior to the coronavirus crisis, many people critiqued and complained about the system of for-profit healthcare and Big Pharma. (See, for example, *Deadly Medicines and Organised Crime* by Peter Gøtzsche.[7])

Furthermore, the data around death counts—perhaps the most elementary variable among all the data—proved to be anything but unambiguous. About 95 percent of registered COVID-19 deaths showed one or more underlying conditions. According to the US Centers for Disease Control and Prevention (CDC), barely 6 percent of deaths account for those whose only condition was COVID-19.[8] In addition, coronavirus victims were usually of advanced age, on average eighty-three years old in Belgium during the first wave, slightly *older* than the average life expectancy. It's a good question: How do you determine who dies "from" COVID-19? If someone who is old and in poor health "gets the coronavirus" and dies, did that person then die "from" the virus? Did the last drop in the bucket cause it to spill over more so than the first?

* * *

All of this is to say that the basic numbers in the coronavirus crisis are not objective data; they are constructed on the basis of subjective assumptions and agreements. Depending on how those agreements are made, the numbers can differ by a factor of no less than fifteen or even twenty. In this "forest of subjectivity," everybody, whether consciously or unconsciously, follows their own prejudices and usually opts for numbers that support their own subjective beliefs. Therefore, some people conclude from the numbers that we are dealing with a problem at the magnitude of the Spanish flu, while others believe that there is nothing especially unusual going on. And these two opposing opinions can, in fact, both be underpinned by "objective data."

The numbers of the dominant coronavirus narrative tend to highly overestimate the danger of the virus. And this tendency is also reflected in the epidemiological models on which the dominant narrative is based. The choice for the lockdown strategy was mainly based on the models developed at Imperial College London. Those models predicted 40 million deaths worldwide by the end of May 2020 if far-reaching measures were not taken to contain the pandemic. Several renowned researchers—for example, Michael Levitt, Nobel laureate in Chemistry; and John Ioannidis, a legend in medical statistics—protested vehemently. They pointed out that the models of the Imperial College were based on wrong assumptions and greatly overestimated the danger of the virus.

By the end of May 2020, it was utterly clear that these critics were correct. None of the countries, whether they went into lockdown or not, came even close to the death count predicted by the models. Sweden is perhaps the most interesting example. This country, according to Imperial College models, would have 80,000 deaths by the end of May if it didn't go into lockdown, which of course it didn't. Its death count was 6,000. And to reach this figure of 6,000 required the "enthusiastic" counting methods described above. Otherwise, it could even have been much less.

The interesting thing is that you'd expect the public narrative and measures to be adjusted (in this case, the introduction of more lenient measures) as soon as the models they're based on are proven incorrect beyond doubt. But that's not what happened at all. Neither public health officials nor the population dialed it back. Something caused society to collectively continue reacting in the same, frenetic way, *as if it were acting out a pressing, psychological need.* In chapter 6, we will discuss this psychological phenomenon.

The limited reliability of the basic data—the number of infections, hospitalizations, and deaths—had an impact on other epidemiological statistics, as well. The infection fatality rate (IFR), case fatality rate (CFR), mortality rate, positivity ratio, and reproduction number—they are all based on these basic numbers. If these numbers vary by a factor of twenty, the statistics based on them will also vary by the same factor. In other words, the epidemiological-statistical discourse sounds sophisticated and looks impressive with its acronyms, calculations to four decimal places, and mathematical modeling of the course of the pandemic, but it is mostly an impressive demonstration of fake accuracy and pseudo-objectivity.

* * *

Some people will object and argue that numbers cannot be relativized onto infinity. Indeed, at some points numbers are open to discussion, but there are matters that cannot be doubted, matters that unequivocally prove the danger of the virus and the usefulness of the measures—don't you think so?

For example, the ICUs are clearly overloaded by COVID-19 patients, aren't they? That's correct. But the way in which we should interpret that fact is another question. Rather than an indication of COVID-19's exceptional danger, the overload seems to be the result of two colliding trends over recent decades: 1. A sharp rise in susceptibility to developing serious symptoms in viral lung diseases in a major part of the population (especially people suffering from obesity and diabetes); and 2. Systematic reduction in ICU beds. The upward trend in the number of patients at risk and the downward trend in the number of ICU beds inevitably had to cross sooner or later. As a matter of fact, this intersection occurred years ago, long before the coronavirus outbreak. The ICU overload has also occurred during recent flu epidemics, for example, resulting in delayed treatments and procedures during those times as well.

So the burden on hospitals can be interpreted as proof of the virus's extreme threat, but it can equally be interpreted as a symptom of inadequate management (progressive reduction of hospital beds), or as a result of declining health (high obesity and diabetes),[9] or as the result of the coronavirus measures themselves (that is, an influx of anxious people, increase in psychosomatic complaints). Depending on the interpretation, radically divergent policies need to be followed.

And yet another remarkable fact: While the limited capacity of the ICUs was the first and main reason for the introduction of the drastic and, from an economic and psychological point of view, extremely destructive measures, no additional ICU beds were created during the crisis. There were no attempts whatsoever to do so. As with individuals, societies also seem to derive some "disease gain" from their psychological symptoms and therefore angle to maintain those symptoms.

Furthermore, the severe lung symptoms associated with COVID-19 in some patients seems to halt any discussion about data. There can be little doubt that those symptoms are real. But how much more severe they are than the symptoms of a normal flu is difficult to determine. There were hardly any lung scans of flu patients, which makes it hard to compare. And in those cases where the comparison was drawn, it sometimes produced unexpected results. At the end of 2020, a study was published that gathered the rare lung scans from flu patients around the world and compared them with lung scans from COVID-19 patients.[10] The study concluded that there was no significant difference. It's hard to say whether this study presents an accurate picture. Since the replication crisis (see chapter 1), we know we cannot assume any study is carefully conducted or that the results present an accurate picture. Furthermore, it is highly probable that the coronavirus has a particularly bad effect on the lungs, based on testimonials from healthcare staff and patients.

The third factor commonly regarded as unshakable evidence of COVID-19's severity is excess mortality. The numbers regarding infections, hospital admissions, and deaths may be subjective, but at the end of the day we can also just check whether there were more deaths during the coronavirus crisis than in prior years. Unfortunately, while this might seem like the most objective measure, there is an intrinsically subjective nature to these data as well, which has also been ignored. As the Ghent University psychologist and statistician Els Ooms showed, excess mortality can be calculated in many ways.[11] For instance, differences in the reference period (the period to which one compares death rates) alone can lead to substantial differences in determining excess mortality.

And after the excess mortality data have been gathered, there is a more difficult task: interpreting these data. Excess mortality is not necessarily an indicator of virus mortality. It may also be a consequence of collateral damage of the coronavirus mitigation measures themselves (reduced immunity, delayed treatment, suicide, depression, addiction, poverty, starvation, and so on), or possibly even the result of the treatment. For example, in 2020, thousands of elderly people in Dutch residential care settings died due to loneliness and neglect during the lockdowns.[12] And a German study suggested that about half of the high mortality in ICUs during the first wave was due to mass intubation (ventilation).[13] It is difficult to say whether these numbers are entirely accurate, but we do know hospitals backpedaled on this protocol in mid-2020 due to its counterproductiveness. It is an important question we need to ask ourselves: What would the virus mortality graphs look like if they were adjusted for these factors?

The following might be the most inconvenient truth of the crisis: that we have called the misery that has been so dramatized in the mass media down on ourselves to a large extent; that the remedy itself has become a significant part of the problem. At the very beginning, in March 2020, I wrote in an opinion piece that fear arises only to a limited extent from real dangers, but that it, in any case, does *create* real dangers.[14] Radical isolation of the elderly and the use of invasive ventilation for ICU patients are probably prime examples.

Vaccination might belong in the same category. All around the world, a decision was made to proceed with a type of vaccine that has been researched to only a limited extent, or at least, the effects of which have been investigated much less thoroughly and for a much shorter period of time than other vaccines. In this, too, we can see that the numbers raise many questions, regarding both effectiveness and side effects. The dominant narrative draws a predominantly positive picture, but out of the enormous flow of data we could just as easily select numbers that draw a predominantly negative picture. Who has heard in the media about the Harvard University study that found no difference in the course of the pandemic between countries with high and low vaccination rates?[15] Who has heard in the media about the study that found the miscarriage rate in vaccinated pregnant women is eight times higher than normal?[16] We are not sure whether these studies paint an accurate picture. But we also don't know whether the numbers that are presented in the media and that confirm the dominant coronavirus narrative do so. Stories make the numbers, rather than the other way around. That's what is at issue.

* * *

Just like that we've arrived at another flaw in the numerical approach to the coronavirus crisis: It largely ignored the collateral damage of the measures, despite them being a crucial factor. There have been hardly any publicly available data and statistics on the number of victims of delayed treatment, suicide, vaccination, food insecurity, and economic disruption. This is all the more remarkable since, from the beginning of the crisis, scientific articles and press releases appeared on a regular basis, pointing out those risks.[17] At the start of the first lockdown, Oxfam, the WHO, and the UN were already warning that death from malnutrition and starvation as a result of lockdowns in developing nations would probably exceed the death count attributable to the virus, even in the worst-case scenario if no measures were taken at all.[18]

The same remarkable disregard could be observed around the mathematical models built to map the course of the crisis. A mathematical model that, apart from the possible victims of the virus, would also represent the possible victims of the coronavirus measures had never been built. When the experts who had built some of the models were asked during their testimony before the British House of Commons why they had not included the collateral damage of the measures in their models, they replied, disarmingly honestly, that this was beyond their expertise as epidemiologists. It was not their job to quantify and draw attention to the collateral damage.[19] This not only shows the limits of the expert and specialist model, in it we can also ascertain a remarkable psychological blindness. And so we see that an entire society can completely ignore what is undoubtedly the most basic question in medicine: Are we sure that the cure is not worse than the disease? In chapter 6, we will see that this narrowing of the field of attention is an effect of the social-psychological process of mass formation.

Furthermore, surprisingly little attention was paid to evaluating the effectiveness of the draconian measures. To the extent that it did receive attention, it underscored that the interpretation of numbers is far from unambiguous. Perhaps, the case of Sweden—the country that, unlike almost all other Western European countries, chose not to go into lockdown and took generally milder measures—provides the best illustration. First, the mainstream media compared the number of deaths in Sweden with countries such as Belgium and the Netherlands. Sweden had fewer victims and therefore, the talking heads concluded, strict measures seemed futile. Then they started to compare Sweden with its neighbors, Norway and Finland, assuming that these two countries had imposed the "normal," stricter measures. Sweden had more than twice as many victims as Norway and Finland, so the talking heads concluded that strict measures were indeed useful. Subsequently, a study appeared, stating that the measures in Norway and Finland had been misjudged: They were actually more lenient than the measures in effect in Sweden.[20] So the conclusion changed once again in the other direction: Strict measures were futile, after all. Whether that will be the final conclusion remains to be seen. What is certain, however, is that, once more, the numbers can be easily adapted to opposing stories.

Comparisons within the United States present us with the same problem. Those comparisons show hardly any difference in absolute numbers of coronavirus victims between the twenty-five states that imposed the strictest measures and the twenty-five states that imposed the most lenient. More or less at the same time, however, a comparison between the ten strictest and the ten most lenient states did show a difference in favor of the strictest states. The story reported in the media interprets the numbers in favor of the dominant narrative, without holding back. If a state that imposed few measures has few victims, it was almost always attributed to an external factor (such as climate or sparse population). Such a state was lucky. If a state that imposed strict measures had many victims, it was also attributed to external factors. Such a state was unlucky, it was hit exceptionally hard by the virus. However, if a state that imposed few measures had many victims, then that was its own fault. It should have taken more measures! And if a state that imposed strict measures reported few casualties, it was reaping the benefits of its decisiveness. In other words, however it turns out, within the dominant narrative, the dominant narrative is always correct.

In addition to country-to-country comparisons, there are also various analyses of the infection curves against the introduction of various measures: the introduction of masking, the start of social distancing, the introduction of lockdowns, the rollout of vaccination campaigns. When such analyses are presented by proponents of the dominant narrative, they usually show that the curve responds immediately to the measures and that infections go down after implementation. However, when the same analyses are performed by coronavirus-critical researchers, they usually conclude that the curve is in no way affected by the measures.

Maybe you think all this applies to information in the popular media but not to articles in high-quality scientific journals? Alas. Whether it concerns the origin of the virus (bat or laboratory), the efficacy of hydroxychloroquine, the (side) effects of vaccines, the usefulness of face masks, the validity of the PCR test, transmissability among schoolchildren, or the effectiveness of the Swedish approach, scientific studies lead to the most conflicting conclusions.

The German philosopher Werner Heisenberg was awarded the Nobel Prize for his uncertainty principle—"It's not a matter that we're not yet sure now; the point is that we can never be sure"—but we don't like it. If the data don't provide certainty yet, we'll collect more. In this way, as a society, we are mesmerized by an endless procession of numbers and never arrive at what really matters: an open debate about the subjective and ideological frameworks from which we interpret the numbers. It is the unspoken tensions, fears, and disagreements on an ideological level that prevent the numbers from settling down and that makes society polarize. The real questions to be asked are situated at the ideological level. For instance: Do we view man as a biochemical machine that has to be technologically monitored and pharmaceutically adjusted, or as a being that finds its destination in mystical resonance with the Other and with the eternal language of nature?

* * *

This chapter opened with some simple examples that challenge a naïve belief in the objectivity of numbers. The Great Britain border measurement example (see figure 4.1, page 50) showed that measurements are always relative and dependent on the measurement unit used; Simpson's paradox demonstrates that even simple, accurate numbers can lead to opposite interpretations. What applies to these simple numbers, applies a fortiori to the frenzied dance of numbers in the coronavirus crisis: Everyone can select numbers that match their own prejudices, everyone can interpret them in such a way that they support their subjective ideological fiction. The almost irresistible illusion that numbers represent facts ensures that people become increasingly convinced that their own fiction is reality.

The use of numbers in this crisis makes us barely realize that what we do respond to are not so much the facts but the stories constructed around facts. Those stories are spun by healthcare workers who genuinely do their best to help, by people who don't want to see their families suffer, by politicians who want to make the right decisions, by academics who want to provide information as objectively as possible. However, they are also constructed by politicians who are under the pressure of public opinion and feel compelled to act decisively, by leaders who have lost control and see their opportunity to take back the reins, by experts who have to hide their ignorance, by academics who see a chance to assert themselves, by man's inherent propensity for hysteria and drama, by pharmaceutical companies that smell dollar bills, by media that thrive on sensational stories, by ideologies that see in a technocratic totalitarian system the only solution to the seemingly insoluble problems of our time.

The influence of subjectivity in constructing and interpreting numbers is so strong that even scientists, whose profession it is to be objective, fall prey to it as well. For example, it's known that in psychotherapy research results usually confirm the researcher's subjective preferences. A psychoanalyst typically concludes from this research that psychoanalysis is the most effective discipline, a behavioral therapist concludes that behavioral therapy is the best therapy, a systems therapist observes that systemic therapy is preferable. This is commonly referred to as the *allegiance effect*—the effect of a researcher's loyalty to a particular theory.

And to be perfectly clear: That effect also manifests itself in strictly controlled, experimental research and also in other scientific domains, such as research into the effectiveness of pharmaceutical medicines.

Most interestingly, this effect manifests itself largely without researchers realizing it. Like hikers on the road without a map or compass, they walk in a circle and return to the point of departure: their own subjective prejudices. That is, of course, a serious problem since the aim of science is to make objective assessments and to exclude subjective preferences from having an impact on the conclusions drawn.

How is it possible for researchers to fall prey to their subjective prejudices? The explanation can be found, in part, among the following issues: Every research procedure requires countless choices, for which there are no strictly logical grounds. Which measuring instruments will I use? How will I interpret the measurements? How do I deal with missing data? And so on. From this vast array of possibilities, researchers unconsciously choose options that will ensure the results they deem desirable.

The fanatical belief in the objectivity of measurements and numbers, which is typical of the mechanistic ideology, is not only unfounded, it is also dangerous. There arises a kind of mutual reinforcement between subjective biases and numbers: The stronger the biases, the more one selects the numbers that confirm these biases. And the more the numbers confirm the biases, the stronger the biases subsequently become. Applied to the coronavirus crisis: A society saturated with fear and unease selects from the myriad of numbers those that confirm its fear. The chosen numbers then reinforce the fear.

As a result, people react in a disproportionate way with all the ensuing consequences: from an economic viewpoint, the recession and the bankruptcy of countless companies and small businesses; from a social viewpoint, permanent damage to the (physical) bond among people; from a psychological viewpoint, even more fear and depression; and yes, from a physical viewpoint, a collapse of immunity and physical health (see chapter 10) as a result of the stressful psychological and social predicament. And we might add: from a political viewpoint, the rise of the totalitarian state. Indeed, if you're convinced that your own subjective fiction is reality, you will also think your reality is superior to the fiction of others. This is how we become convinced that our fiction can be imposed on the other by any possible means.

At the beginning of the chapter, we described that the mechanistic ideology aims to instate a technocratic society that is governed on the basis of "objective," numerical information and in which subjective preferences and abuse of power are eliminated. But at the end of this chapter, we conclude that naïve belief in the objectivity of numbers leads to the exact opposite. The dominant ideology repeatedly presents numbers in the mass media that confirm its own narrative, resulting in a largely fictitious reality in which a large part of the population firmly believes. The perception of reality is determined time and again by numbers that, a few months later, turn out to be very relative, sometimes plainly wrong, or even deceptive. But in the meantime, these numbers are used over and over to impose the most far-reaching measures and to set aside all basic tenets of humanity: Alternative voices are stigmatized by a veritable Ministry of Truth, crowded with "fact-checkers"; freedom of speech is curtailed by censorship and self-censorship; people's right to self-determination is infringed upon by imposed vaccination, which imposes almost unthinkable social exclusion and segregation upon society.

The discourse surrounding the coronavirus crisis shows characteristics that are typical of the type of discourse that led to the emergence of the totalitarian regimes of the twentieth century: the excessive use of numbers and statistics that show a "radical contempt for the facts,"[21] the blurring of the line between fact and fiction,[22] and a fanatical ideological belief that justifies deception and manipulation and ultimately transgresses all ethical boundaries.[23] We will describe these characteristics in detail in chapters 6 and 7. But first, in chapter 5, we consider the social conditions that prime a society to cling to this numerical illusion of certainty. We will see that the flight into false security is a logical consequence of the psychological inability to deal with uncertainty and risk, an inability that has been building up in society for decades, perhaps even centuries.

CHAPTER 5

The Desire for a Master

In previous chapters, we discuss how science tipped from open-mindedness to dogma and blind conviction (chapter 1), how its practical applications isolate people from one another and from nature (chapter 2), how its utopian pursuit of an artificial and rationally controllable universe equates to the destruction of the essence of life (chapter 3), and how its belief in objectivity and measurability of the world leads to absurd arbitrariness and subjectivity (chapter 4). In this chapter, we will discuss the fate of another great ambition of science: to liberate man from his anxiety and insecurity and his moral commandments and prohibitions.

Religious discourse for centuries darkened the human soul with irrational fear of hell and damnation. Suffering and disease were God's punishment, aging and infirmity were something to be accepted, carnal pleasures were tarnished with the stigma of sin, society was suffocated with sullen commandments and prohibitions.

Sometime during the seventeenth century, the star of the human intellect appeared in the sky. Man started to look outward; neither God nor devil appeared before his rational eye. The fear instilled by the religious discourse was declared unfounded; there was no longer any reason to accept the social contract imposed on society by the clergy. Man started to explore the world that surrounded him, studied the human body and the causes of disease and suffering. The human condition was not to be accepted—it had to be *improved*. For three centuries, an energetic optimism prevailed. The human condition could be made enjoyable. Disease and suffering would be cured by the power of the human intellect.

The commandments and prohibitions of the past were declared superfluous, unnecessary to steer society in the right direction. An increasingly loose morality would eventually reconcile man with carnal desires, formerly perceived as threatening. The crippling censorship of anything contrary to the religious discourse disappeared. Freedom of speech became a basic right, education became universally available, legal assistance became a right for all, love was stripped of its duty to marry and have children, sexuality was restored and its coupling with sin and corruption was undone.

Somehow, however, this process turned in the opposite direction. The idealization of the human intellect eventually led to an intensification of fear of disease and suffering, while interhuman relationships were marked by uncertainty and confusion. The old commandments and prohibitions were eventually replaced by a jungle of rules and regulations and a new, hyper-strict morality. How can we understand this from a psychological perspective?

* * *

No matter how much the knowledge of the mechanistic aspects of the human body increased and how much money was spent on health care (which, in Western European countries, easily exceeds 10 percent of the gross national product), the fear of disease and suffering did not disappear at all. Headlines in recent years have left no doubt about it: It is irresponsible to send teenagers to school on a moped,[1] swimming in rivers or ponds in hot weather is not recommended due to the risk of bacterial contamination,[2] oral sex may cause throat cancer,[3] shaking hands is too dangerous because of virus transmission,[4] yes, even sitting next to a smoker who is not smoking may cause harm to your health.[5]

These are just a few of the endless stream of media reports that illustrate how much twenty-first-century people's lives are dominated by fear of physical adversity.

Suffering is by definition unpleasant, but there have been times when people were more resilient to it. In the seventeenth century, when Jesuits tried to convert Native Americans to Christianity by burning them at the stake, the missionaries discovered, to their great frustration, that the Indigenous people were unimpressed. Over time, the Native Americans themselves suggested other, much more painful forms of torture. "Why always at the stake?" they asked the missionaries.[6]

Not only has the thought of physical suffering become more unbearable, people have increasingly become less risk tolerant. The insurance mania that spread over the last few centuries is perhaps the best illustration. It started meritoriously during the nineteenth and twentieth centuries when accident and fire insurances gradually became established and institutionalized. Then it expanded to life, hospital, travel, and cancellation insurances, and ultimately to insurance for just about everything. Today, not only trees, plants, dogs and cats,[7] but also Christiano Ronaldo's legs, Jennifer Lopez's bum, Taylor Swift's breasts, Julia Roberts's smile, and David Lee Roth's sperm have been insured against damage for up to millions of dollars.[8] Not to mention the insurance against heartbreak, meteorite impacts, and damage caused by spirits and ghosts and alien abduction.[9] It should come as no surprise that nowadays, you can also insure your insurance (with Lloyd's of London, for example).

Desperate attempts to avoid any risk take their toll, however, and not only in terms of insurance premiums. Medical interventions, which should eliminate suffering, are increasingly a source of despair themselves. The widespread consumption of psychotropic drugs, painkillers, and other pharmaceutical products has led to tens of millions of addicts and countless deaths. Screening for cancer and other diseases is not only harmful in and of itself but also leads to ever more unnecessary, harmful interventions, such as unnecessary breast amputations and side effects of chemotherapy.[10] In addition, preventive medicine threatens to render life sterile and inhumane. The COVID-19 response is a good example: Maniacal avoidance of infections led to an incalculable increase in suffering due to delayed treatments, domestic violence, psychological despair, and food insecurity in the developing world.[11] In other words, frantically trying to avoid any danger has, paradoxically, become very dangerous.

The effects of this desperate attempt to control life go beyond a detrimental impact on our physical health. It also severely affects our freedom and rights, as individuals. At the beginning of the twenty-first century, the War on Terror, for example, led to a serious violation of privacy. In fact, it was part of an ongoing and growing endeavor to attempt to control and isolate "dangerous elements" in society. The tradition of the Enlightenment led unintentionally to what Foucault called *le grand renfermement*: More and more "dangerous" groups were imprisoned.[12] In the nineteenth century, it affected "only" psychiatric patients, prostitutes, and criminals; in the twenty-first century, it affects just about everything and everyone. Animals are caged because of the bird flu, the global population is placed under house arrest because of the coronavirus. Humans and animals—potential spreaders of disease—are too dangerous to each other to be let loose.

* * *

The societal increase of fear and insecurity leads to two other psychological phenomena: narcissism and something I call *regulation mania*. In order to understand this connection, we need another piece of developmental psychology. We will start by explaining the connection between human insecurity and narcissism.

In chapter 3, when we discussed the difference between digital and "real" conversations, we described how an infant resonates symbiotically with its mother through the early exchange of body language and, in this way, realizes the primal desire for blending with the Other. There is a lack, however, in this early paradise. In a certain sense, the child hardly exists there as a separate psychological being. During the first months of life, before it can recognize itself in the mirror, a child cannot form a mental-visual image of its own body. As a result, it does not know where its body ends and where the surrounding world begins, and it situates its own sensations not only in its own body but also in the people and objects surrounding it (animism). A concrete example: When it gets a jab in its arm, it doesn't look at its arm because it does not realize that the pain sensation is located there. And the reverse is also true: The child feels the sensations of others directly in its own body. For example, when it looks at someone being beaten, its face shows the same grimace and it cries as if it were being beaten itself (transitivism).

In this symbiotic but also chaotic amalgamate of experience, the child has to mentally grasp what is at the core of its existence: It has to find out, through interactions with the mother figure, what it needs to do to ensure her care and closeness. At this point, it is interesting to make the comparison with a young animal. Young animals and mammals are also dependent on their mothers, and they also try to ensure themselves of her care. However, there is a crucial psychological difference with the human child, situated at the level of the communication system.

An animal establishes the bond with another animal through the exchange of signs. The signs—typical cries, postures, movements—have a well-established connection to their point of reference. One sign refers to danger, another sign indicates that food is on its way, yet other signs indicate sexual availability, submission, or dominance. Whether an animal sign system is simple or complex, whether its mastery is innate or passed on from generation to generation through learning, the signs are generally experienced by the animal as unambiguous and self-evident. Their exchange can lead to a fierce battle under certain circumstances—for example, the battle between male sticklebacks when their red bellies indicate that they want to reproduce—but it will not usually lead to persistent doubt or uncertainty.

In humans, this is different. Human communication is full of ambiguities, misunderstandings, and doubts. This has all to do with the following: The signs—or more correctly, the symbols—of human language can refer to an infinite number of things, depending on context. For example: The sound image *sun* refers to something completely different in the sound sequence *sunshine* than in the sound sequence *sundering*. Therefore, each word only acquires meaning through another word (or series of words). Furthermore, that other word, in its turn, also needs another word to acquire meaning. And so on to infinity. There is always a word missing to definitively capture the meaning of words. For this reason, language as a rational system—as a system in which words acquire meaning axiomatically—has an intrinsic, irreparable lack. This immediately makes clear that even the insurance-of-the-insurance cannot free man from his linguistic uncertainty.

This has direct consequences for interpersonal interactions. We, as human beings, can never convey our message unambiguously, and the other can never determine its definitive meaning. It goes even further: We don't even really know our own message. We never know exactly what we want to say, simply because our thoughts also work with words and so there is always a word lacking on that level, too. That's the reason why we so often have to search for words, so often struggle with saying what we really want to say, so often feels like we're saying something we didn't really want to say or that we meant something slightly different. There is no trace of this in the animal world: Their communicative behavior does not show these hesitations and stammerings.

We tend to think that humans distinguish themselves from animals by *greater* knowledge and awareness, but the most typical difference is that, unlike animals, we are almost constantly tormented by a *lack* of knowledge. Therefore, the central questions in a human's life, those that relate to his position in the desire of the Other, never receive a definitive answer. What does the Other think about me? Does he love me? Does he find me attractive? Do I mean something to her? What does the Other expect from me? What does he want from me? It is around these questions that every human encounter and, by extension, the whole of human existence, gravitates. There is no indication whatsoever of this in the animal world: You will never see an animal sitting on a couch worrying about the meaning of its life or about what it means to another animal.

This indefiniteness of the human world of symbols has, a bit surprisingly, been going on from the very beginning of a human life, at a time when language is still rudimentary and does not yet refer to objects.

The great French developmental psychologist Henri Wallon noted that from the very beginning, you see something on the faces of children who interact with their caregivers that you do not see in any other living being. When a newborn child fixates and imitates the facial expressions of the mother, its face already expresses a subtle sentiment of *question*, as if, even at this very early stage of its existence, it is confronted with something that is missing in the form language of the Other.

Therefore, a human child is, in contrast to a young animal, in a state of deep uncertainty about its mother's messaging. And that makes it difficult to gain mental control over her. What does she want from me? What should I do to ensure her presence? However undifferentiated the mental system may be at that moment, these questions do arise even in these earliest months of life. This explains one of the most curious phenomena that occur in a child's development. Around six to nine months old, a child recognizes itself in the mirror for the first time, usually while the mother points enthusiastically at the mirror image. That in itself is not unique to humans; dolphins and higher monkey species are also able to do this without problems. However, as Charles Darwin noted, the recognition in a human child is accompanied by something that does not occur in any other animal: The child cheers with joy.

What makes that recognition in the mirror so pleasing, while it leaves other animals completely indifferent? Unlike an animal, the human child suffers from a constant tension due to the eternal elusiveness of the world of symbols, in which it is immersed from the first moments of its existence. And this applies in particular with regard to the most central question: What does my mother want from me? That tension is instantly removed when he sees there, before his very eyes, a mirror image with which it coincides and at which the mother is pointing with great enthusiasm. This reflection instantaneously tells the child who it is and needs to be in order to be the object of the mother's desire. That image in the mirror seems, all at once and in all its concreteness, to offer an answer that language never can: I am *it* for the Other. This experience is the archetype of the narcissistic experience. It is so overwhelming that some people obsessively look for such experience later in life in an effort to avoid the feeling of lack and insecurity in human relationships.

However, this experience also takes a toll, both for the relationship and the individual. In order to avoid the re-emergence of the underlying insecurity, the child has to engage in an aggressive rivalry with everyone else that also draws the mother's (later, a love object's) attention: Only one person can be the mother's object. The more one chooses to master insecurity through identification with the mirror image, the more he has to outperform, belittle, and even destroy others—basically the more he loses his humanity.

Additionally, this dehumanization is reinforced by the fact that identification with one's own mirror image reduces the capacity for empathy. This identification provides the child, for the first time, with a global visual picture (or a substitute for that in blind children) of its own body. This global image enables the child, for the first time, to draw a boundary—literally, a mental line—around its body. To a certain extent, this is necessary to build a stable Ego structure. Without such an image, the child cannot mentally experience itself as a unit. However, in excessive narcissism, the mental-visual boundary between the subject and the Other becomes so thick and pronounced that the subject becomes mentally locked up in this self-image. The visual self-image then attracts the mental energy and attention to such an extent that the image of the Other no longer "lights up" in the mental experience. As a result, one can no longer feel affinity or empathize with the other person or the world. In other words: Excessive narcissism comes at the expense of empathy. To the extent that it diminishes a person's ability to resonate with others and with the world, it renders that person lonely and isolated.

From this line of reasoning, we conclude that excessive investment in the mirror image is an overcompensation for the uncertainty that human language generates in interpersonal relationships. But in the extreme, this overcompensation is always a fallacious solution. One tries to assure oneself of symbiosis with the Other, but ends up in psychological isolation from and destruction of the Other. And also in *self* destruction. It's best to imagine this in a concrete-visual way: All energy that is inside the psychological system is sucked away and invested in the surface of the body, meaning in the visual image of the body. It is no coincidence that people who focus heavily on appearances often say that they feel "empty" during psychotherapy sessions.

In recent decades, we have seen that, along with an increase in fear and insecurity, narcissism is also on the rise. It has become a cliché to say that our society focuses more and more on external ideals, but there is unmistakably something true about it. The number of surgical procedures that "fix" the body in order to resemble a social ideal is rapidly increasing, the sale of steroid and protein cocktails to force the body machine into a visual ideal has grown spectacularly, taking selfies forms part of the established repertoire of (a)social behavior, houses and gardens resemble staged photos from home decor magazines, commercials and billboards present stylized ideals of cars, haircuts, and clothing. In essence, this trend boils down to a growing obsession with fallacious visual "solutions" in an attempt to eliminate the irresolvable uncertainties in human relationships. At the same time, we naturally also see a sharp increase in the psychological phenomena associated with excessive investment in the outer ideal image: experiences of loneliness and inner emptiness, and of feeling consumed by an exhausting competition with others (the so-called rat race).

* * *

In addition to narcissism, there is a second social phenomenon that is directly linked to the increase in fear and insecurity: the enormous increase in the number of rules, sometimes referred to as *regulitis*. We can situate this regulation mania very simply within the same developmental psychology that I described above.

Recognition of its own mirror image ensures that the child is able to psychologically demarcate its own being (body) from the surrounding world. It is only at this point that external objects mentally begin to exist for the child. This causes the function of language to change. The words now begin to refer to those external objects (they take on a referential function) and thus also acquire meaning. Previously, this was hardly the case. Prior to the "mirror moment," the child's expressions were mainly physical, instinctive "acts" that expressed bodily sensations to realize a symbiotic resonance with the Other.

The moment that words acquire meaning, the relationship with the Other will also be raised to another level. The child now obsessively tries to understand the words that the other person uses to express his desires. What exactly does it mean to be "good"? What do I have to do to be "a brave girl"? Simply put, it wants to know the rules it must follow in order to be loved. At certain moments, this takes the form of a demand for rules; no matter how well a rule is defined, it is still too unclear and requires additional definition. And since the words in which the rules are formulated acquire meaning only by means of other words, the child starts to wonder about the meaning of every possible word.

Around the age of three and a half, this obsession with the meaning of words culminates in the so-called "why" phase. In this phase, a child endlessly poses "why" questions." "Why is this a donkey?" "Because he is braying." "Why is he braying?" "Because he is angry." "Why is he angry?" And so on. In this stage, the child sees the parent as an omniscient master, and despite the fact that he sometimes resists submission with extreme stubbornness, he also demands that the parent takes that position. He has to know everything. If the parent cannot determine what she wants, the child doesn't know how to comply with her desire. That's the point at which the child is confronted with the human primal insecurity and is overtaken by the human primal fear: being left behind by the Other (primarily by the mother) because it is unloved.

The child's attempts to make the rules unambiguous and conclusive are doomed to failure because, again, human language can never acquire definitive meaning. The more persistently the child tries to make the rules unambiguous by questioning the parents, the more he inevitably loses himself in complex and contradictory interpretations. In children with a compulsive disposition, this happens clearly, and they end up in almost complete inhibition, entangled in an endless pursuit of mental perfection that gets more and more bogged down. We will see later that children are eventually liberated from their demand for rules by accepting that a definitive answer to the question with respect to the desire does not exist. This, at the same time, will require that they give up the narcissistic striving to be the object of the Other (which usually is, at that stage, the mother).

* * *

This developmental psychology can also be applied at a social level. Society is—it's hard to ignore—increasingly bogged down in an endless proliferation of rules. On the one hand, such rules are imposed by the government, but on the other hand, there is also a call for more rules—a hyper-strict morality—from the population itself. Like narcissism, this is a frantic attempt to contain the surge of fear and insecurity in human relationships.

It is indeed a striking phenomenon: Since the beginning of the twenty-first century, a new morality has arisen from the belly of Enlightenment thinking, which in a number of respects is stricter, more vagarious, more irrational, and more hypocritical than the prior religious morality, which the Enlightenment sought to obliterate in order to set people free. With the rise of the woke culture, society fell prey to implicit and explicit rules that made every detail of human interaction more precarious. In the wake of the #MeToo movement, students were taught how to flirt legally and compliantly,[13] freshmen initiations were subjected to increasingly strict regulations,[14] Sweden introduced a law stating that sex is only legal if the parties involved give their consent in advance via a signed contract,[15] nude figures of the paintings by Flemish masters were no longer allowed to be posted on social media,[16] and Netflix introduced a rule stipulating that eye contact between employees should not last longer than five seconds and that employees are not allowed to ask for each other's phone numbers without asking permission for asking first (!).[17] The new norm has become so stringent that even suggesting that there is a physical difference between a man and a woman can be considered a violation of sexual integrity.[18]

The Black Lives Matter movement is captured in this trend as well. The tendency toward increasingly exhaustive standards with respect to racism intensified to little productive end: The chances that such rules truly contribute to the overcoming of the narcissistic superiority feelings that are involved in racism is, in fact, rather small.

The climate movement has also given rise to a new category of crimes: environmental. To the point that using a wood-burning stove, eating meat, or living off-grid in the countryside are considered environmental violations, environmental ideology has been taken to such an extreme that it has become opposed to that which it originally aspired to: getting back to nature. Environmental violations are also rather selective and inconsistent in their strictness. For example, reducing one's carbon footprint is taken to extremes, while there is remarkable leniency regarding energy consumption through internet use (which is as high as the energy consumption from all air traffic combined) and the "mining" of Bitcoins (which is as high as the energy consumption of an average Western European country). And also the environmental damage caused by mining ores for batteries for electric cars is rarely discussed. The environmental movement was once a dissident voice, but with its turn toward "ecomodernism," it has clearly merged into the dominant mechanistic ideology.

This regulation mania is also directly visible in the public space. My office at Ghent University looks out over a major intersection. Over the past twenty years, I have watched this intersection change from a large asphalt plain with a few sparse white lines to a mosaic of lines and colored areas, indicating where cyclists, pedestrians, and cars are and are not allowed to go, with ever more traffic signs and traffic lights mounted thereon. And it isn't only the intersections. In train stations, you have to buy a ticket to have access to the toilets, yellow squares indicate where smokers can indulge in their dangerous addiction, and you're allowed to park only in certain delineated—paid—parking spaces for a certain period of time. During the coronavirus crisis, this phenomenon reached its temporary peak with an endless number of arrows indicated on floors and stairs, showing where to walk and in which direction, signs reminding you that you are required to wear a face mask, confined spaces demarcated by crash barriers preventing one bubble from coming into contact with another one at festivals and cultural events, red and green dots on chairs indicating where you are and are not allowed to take a seat in the theatre. The moment at which the rules will be abolished is postponed endlessly and will actually never arrive, if it depends on the proponents of the current coronavirus approach. Indeed, the possibility of a few hundred thousand deaths from a "normal" flu virus would surely justify the introduction of similar measures in the future.

Furthermore, the jungle of rules that are activated in response to all kinds of threats varies from location to location. During the coronavirus crisis, mayors are able to adjust rules in their own jurisdictions at their discretion. And the rules also change over time. During thunderstorms, terrorism, and viruses, they can easily switch between green, yellow, orange, or red codes. In the long run, the rules also become so detailed that one either gets angry or has to laugh: In the summer of 2020, it was ruled that an opening dance would be allowed at weddings, however, not the polonaise.[19] The coronavirus apparently knows something about dancing. Keeping up with the rules proves an impossible task, which puts the competent authorities themselves in a state of hopeless confusion. At a certain point during the second lockdown in 2020, the website for the Belgian Ministry of Health stated that noncohabiting partners were allowed to visit one another, yet the police could still fine people for doing so.

The problems exposed by the New Morality are legitimate. Sexism and racism are symptoms of cultural decline; people have to take care of nature (or the climate) or we will irreparably destroy it, and solidarity with victims of the coronavirus (and victims of the public health response) is evidence of our humanity. This doesn't mean however that the suggested solutions are legitimate. They are excessive, inconsistent, and counterproductive in many respects. In the #MeToo discourse, the lines between clumsy flirtation and rape are blurred; in the Black Lives Matter discourse, to make any reference to skin color is like walking on eggshells; the climate movement alienates man even more from nature; and with the coronavirus crisis, health care has become an attack on life and liberty. Moreover, as Freud pointed out, the repressive nature of the new morality is fueling an exacerbated "return of the repressed": Between 2015 and 2020, the use of sexist language doubled and the use of racist and menacing language tripled on social media.[20] This counterproductivity must be acknowledged, albeit with the reservations we always have with regard to numbers and statistics.

The new morality is also more and more aggressively enforced, both by the government and by the population itself. Support for free speech, freedom of the press, artistic freedom, and basic self-determination is decreasing at an alarming rate: J. K. Rowling was fiercely attacked (to the point of her house being molested) when she scorned a full-woke reference to "people who menstruate" instead of "women";[21] German insurers want an alcohol lock in every new car;[22] the *New York Times* editorial page editor was fired for publishing an op-ed by a right-wing politician about the death of George Floyd;[23] in Australia, a man was declared a public enemy of the worst sort and hunted by the police and army for not complying with the mandatory quarantine after a positive COVID-19 test (which actually may well have been a *false* positive test).[24]

* * *

You could still doubt whether these excessive, absurd, and inconsistent regulations are typical of contemporary society. Were there really fewer rules in the past? And were the rules less absurd in the past? The 613 commandments and prohibitions of Jewish religious regulations (the *halacha*) have been around for thousands of years. They subject the lives of Orthodox Jews to rules down to the smallest detail. And Jews themselves are often the first to admit they are not always logically understandable. In addition to the rules that have a logical basis (the *mishpatim*), there are also those that perpetuate the bond between man and the Eternal and which cannot be logically understood (the *chukim*, which include the dietary laws and circumcision).

Rules were also rampant among Indigenous peoples. Totemic tribal societies often maintain a complex system of rules of conduct, precepts, and taboos that substantially strip everyday life of its spontaneity. Specific objects, such as weapons and clothing, cannot be touched in specific situations, certain foods are prohibited (including the flesh of a totem animal), and even certain footprints cannot be followed (among the natives of Leper Island, for example, brother and sister avoid each other's tracks).[25] And contrary to what a romantic revisionist portrayals of tribal societies might suggest, there is no free love and sexuality in the wilderness. Among certain Australian Aborigines, for example, a given tribe might have historically been divided into twelve clans. Both casual sexual relations and long-term sexual relations were allowed only with members of three specific other clans. Therefore, for a man, three out of four women were already taboo beforehand. Violations of both men and women are punished by no less than death. The Ta-Ta-thi tribe in New South Wales has a somewhat milder history. They killed the man, and the woman was "merely" beaten and impaled on a pole until she was *almost* dead.[26]

The comparison between religious, indigenous, and modern systems of law is far beyond the scope of this book, but there's no doubt that differences exists. For example, both religious and indigenous systems of law were, in general, categorical and as such rather clear. And another important difference: They were also stable. The current modern legal systems are not. They change quickly and unpredictably. If you buy a car in Ghent today, it is possible you will not be allowed to visit a different city next year because your new car will have the wrong Euro standard. Moreover, the rules are constantly increasing in volume. For instance, data show that, proportionally, more and more time and energy is being spent on the formulation, observance, and implementation of all kinds of rules. On a political level, we see how the regulation mania historically advanced through increasingly bureaucratic forms of government, first in the imperialism of the late nineteenth century (as a logical sequel to colonialism, the nature of which, in itself, was not yet bureaucratic), then in the rogue gang-totalitarianism of the first half of the twentieth century (Nazism and Stalinism-style regimes) and subsequently in the rising technocratic totalitarianism of the early twenty-first century. All these state systems were characterized by increasingly complex and absurd regulations.

This change in regulation is also reflected in the spectacular increase of administrative jobs throughout the nineteenth and twentieth centuries. Between 1840 and 2010, jobs in administration, management, and services increased from 20 percent to 80 percent of the total number of jobs.[27] Administrative staff at American universities more than doubled in thirty years.[28] It is not only the number of administrative *jobs*, the number of administrative *tasks* is also increasing, even in professions that by nature have little or nothing to do with administration. Whether shopkeepers, farmers, or teachers, they all have to deal with a growing number of regulations and are forced to spend more and more time on administrative tasks.[29]

* * *

The regulation mania, in all its extravagance and absurdity, undoubtedly contributes to the psychological troubles of our time. The contradiction and ambiguity of so many rules creates a neurotic dog-of-Pavlov effect and its excessive nature takes away the satisfaction, spontaneity, and joy of life. There is less and less space for autonomy and freedom. For example, at first glance, there are only advantages to the so-called "zipper rule" that requires late merging on European roads. However, it constitutes a subtle psychological disadvantage. Enforced late merging removes the personal choice, as well as the possibility of a small but powerful human encounter—a situation in which one person chooses to give priority to another. A driver no longer has the option of acting with spontaneous generosity, because he is obligated to do so. This may seem inconsequential but it isn't. It is precisely those moments of human-to-human encounters that nourish the social bond from within. Without those moments, the social fabric shrivels, and it is only a matter of time until society disintegrates into a loose collection of atomized individuals.

The suffocating effect of an excess of rules is most noticeable when it suddenly disappears, for instance, when you arrive in a small French village and there are no white lines painted on the streets that tell you exactly where to drive and where to park your car. You can park along the road, without paying and for an unlimited period of time. Or a rural train station where you don't have to pay at a parking meter in the parking lot, where the toilets are freely available, and where the platforms are accessible at all times. It is somewhat reminiscent of the buzz of the air conditioner in your office. You don't notice that it bears down on you until it disappears at six o'clock, and you experience a moment of blissful peace.

The over-regulation has mostly advanced without us realizing it. It also exerts its suffocating influence mostly without us realizing it. But every time the regulation machine is tuned up higher, we lose some space for our existence as living, human beings. It creates a kind of vicious circle: In order to reduce unease and frustration in social spaces, we make more rules, protocols, and procedures. Those rules subsequently lead to more discomfort and frustration. We respond to that with even more rules. And each time the regulatory fabric is woven a little more tightly, the human being receives less oxygen. If the trend toward a hyper-regulated society continues, an increase in suicide attempts will be a logical consequence. The euthanasia machine—a box in which you can relieve yourself of life painlessly with helium gas—will be the ultimate consequence of mechanistic thinking.

Regulation mania, as manifested in government bureaucracy, attempts to render social interactions rational and logical by squeezing them into preformed templates. In this respect, the ideal bureaucrat is identical to a computer: They strictly adhere to the logic of their system without being "distracted" by the individuality of the people they "assist." For this reason, a bureaucratic system generates exactly the same frustration as a computer: We are confronted with a mechanical Other who is in no way sensitive to our individuality as human beings. A computer is not so much an unfair or unjust Other; it is an Other who imposes a relentless logic. It doesn't matter if we have to go to a meeting in five minutes and urgently need to print another report—the computer won't be more understanding or lenient ("computer says no"). In this respect, the computer resembles the ideal totalitarian leader: He strictly and ruthlessly imposes his logic on the population. We'll talk about this more in part 2.

* * *

This is why narcissism and regulation mania are fallacious solutions for the uncertainty and fear that language introduces into human relationships. They lead to social isolation and are ultimately self-destructive. But there are also *real* solutions. We return one last time to developmental psychology.

We arrived at the "why" phase, in which a child keeps asking his parents (and sometimes all the adults around him) "why." The result of that persistent questioning is that the child eventually starts to sense something crucial: If it continues to ask "why," the parent eventually has to admit the limitation of his knowledge. It is at this stage, for most children, that the belief that his parents are omniscient and omnipotent comes to an end. After recognizing himself in the mirror, this is the second revolution in psychological development.

From then on, the child intuitively understands that even his authorities do not fully understand the meaning of the words and that the uncertainty can never recede. At that point, there are two possible responses: fear or creativity. To the degree that fear predominates, the child may cling to narcissism and the craving for rules. But the realization of the inevitable also opens up another possibility: Since no one definitively knows the meaning of words—What is "being good," What does it mean to be a "brave girl" and so on—a child can emancipate himself from the discourse of its parents and give its own creative answers to these questions and hence start to realize its own, unique way to live its life.

On the one hand, the child has to seize his opportunity and realize himself creatively in the space that has arisen. On the other hand, the parents play an important role in this process as well. They can confirm and support the child's efforts to, little by little, give meaning to life and make his own choices. Or they may, in overt or more covert ways, try to maintain the status of their omniscience and continue to make choices on behalf of the child. In the first case, the path to individuality will probably be smooth. In the second case, there is a good chance it will encounter crises and storms. It is difficult to predict which of these two scenarios in the end will yield the most original results.

With the realization that the discourse of the parental gods is not entirely accurate, we see a fledgling sensitivity to a discourse that does not *intend* to be completely accurate: fiction and poetry. During this period, the child mainly hungers for stories about parents and grandparents, stories that, in their *Dichtung und wahrheit* (facts and fiction), provide the child with a basis for its identity and principles about how to behave ("a member of our family is polite, works a lot, likes to eat and drink"). Psychologically, these principles differ radically from the rigid rules it relied on before: They are loose guidelines that are followed faithfully but flexibly in every new situation the child is confronted with. It is these principles that free the child from the rampant craving for rules.

The looser use of language and words, not aimed at definitively assigning meaning, allows the child to rediscover something in the unique context in which he finds himself. After the long detour of acquiring a self-image in the mirror stage and the period of nascent rationality, the child finds, in stories and poetry, echoes and scents of the lost maternal paradise of its earliest months of life.

Therefore, the creation of individuality, through the transition from a logical-rational to a evocative-creative use of language, is a third possible response to the fundamental uncertainty of the human condition. That does not equal a fall into irrationality (we will come to this in detail in chapter 9). However, this creative act, in contrast to narcissism and regulation mania, is indeed a real solution to the uncertainty inherent in human relationships and human existence in general. It connects man with the Other and leads to resonance with (love) objects instead of psychological isolation and (self-) destructiveness. At the same time, it also creatively realizes individuality and psychological sovereignty.

* * *

Let us return for a moment to the questions we asked ourselves at the beginning of the chapter. How is it that the Enlightenment tradition led to *more* fear and insecurity and, eventually, hyper-strict morality? Didn't it explicitly aim at the opposite? The developmental psychological scheme, as outlined above, makes the answer quite simple. The Enlightenment tradition, the ideology of Reason, was a persistent attempt to squeeze life into logic and theories. It placed all symbolism, mysticism, fiction, and poetry secondary. But this is exactly the kind of discourse that allows us the ability to respond to the uncertainty of life with creation and individuality and to find words that resonate with the Other.

That's how uncertainty turned into fear, and the only psychological means available to combat that fear were narcissism and an endlessly rampant regulatory discourse. It is especially this second attempt to "solve" the fear that is especially important here. The more we attempt to eliminate the fear and uncertainty through rationality and rules, the more we collide with failure. Remember what we said about the structure of language: The last word, which should remove uncertainty and bring final resolution, does not exist. Both logically (from the developmental perspective, as we discussed in this chapter) and historically (as we'll see in subsequent chapters), it is precisely at this point that man turns to the opposite of what he pursued in his desire for freedom: the absolute master—the totalitarian leader—who claims to have the last word.

This sheds a different light on social phenomena such as #MeToo, Black Lives Matter, the climate movements, and the coronavirus crisis. They are related to real problems, but those problems are not the real reason for the existence for these phenomena. They arise mainly from the pressing need among the population for an authoritarian institution that provides direction to take the burden of freedom and the associated insecurity off their shoulders.[30] And the government is eager to fill that vacancy. Little by little, it limits the individual's freedom of choice and makes choices for him: It imposes tobacco, sugar, and fat taxes; it determines how health and immunity should be pursued (no access to public spaces or the workplace without a vaccine); it determines how much alcohol you can consume when you are in COVID-19 quarantine (six beers a day in Australia); it bans religious symbols from public spaces, and makes the signs of its own ideology mandatory (without a QR code, the doors will remain shut). The individual will eventually lose even the right to make decisions about his own life. When patients report suicidal thoughts, therapists are under pressure to proceed to collocation; suicide is not allowed under any circumstances. However, if the government approves, you can get permission for euthanasia for reason of mental suffering. In other words, from now on, the government determines when you are allowed to die. The educating and disciplining function of the government is becoming more complex every day and, for this reason, an efficient system becomes necessary. At first, a social credit system seemed like something that would only be possible in communist-totalitarian China, but Australia is preparing to introduce a similar system[31] and some municipalities in Belgium are already using their own virtual currency, which you can earn with "exemplary conduct."[32] (I suppose an unelected technocrat will define what that means.) Should we fear that here, too, like in China, people will be placed in re-education camps, based on an Orwellian computer algorithm, if they have collected too many bad points?[33] The government apparatus, impersonal but crafty, has already anticipated that naughty children will demand some space for individuality: It has disarmed the population beforehand and secures a monopoly on violence.

Ultimately, the position of the totalitarian leader is impossible, simply because, despite his megalomaniac faith and ideological fanaticism, he too is subject to the structure of language. He can only *pretend* to have the last word. This last word floats elusively in the resounding spaces of poetry, fiction, and symbolism—that is, in the space of the type of discourse that admits that it is incomplete. The person who still wants to be in the position of the absolute master falls into errors and inconsistencies, and eventually into outright lies and deceit. We have already discussed this phenomenon in chapters 1 and 4 where we talked about the crisis in the sciences, but we see it just as well at the level of public discourse.

The excessive pursuit of transparency and hypercorrectness also tilts in the opposite direction, namely in pretense and deceit. Just look at the media coverage: Government labels of quality products are often unreliable;[34] the government bans pesticides but then sends officials out to explain to farmers how to bypass the tests that can detect these pesticides (as aptly described in Isabelle Saporta's *Vino Business*);[35] the encryption companies from which you buy the software to protect your privacy turn out to be owned by secret services of the government.[36] Even making health care more transparent and correct—one of the main twenty-first century government's priorities—turns out to be the opposite. Electronic patient records are shared en masse without the patient's consent,[37] they are hackable (as happened to tens of thousands of records in Finland),[38] and insurance agents have access to these records.[39]

* * *

That's how the rationalistic approach to life led to an inability to manage fear and uncertainty in a productive way: Narcissism and regulation mania intensified the problem they seemed to solve, resulting in a psychologically exhausted population that craves an absolute master. It paradoxically looks for that master, in accordance with the dominant view of man and the world, in the mechanistic ideology—that is, the ideology that caused the problem to begin with. This is also the ideology that tempts the minds with its immense manipulations of matter and that seems to have the facts on its side with numbers and statistics. It is that condition of the population—fearful, socially atomized, and yearning for direction and authority—that is the perfect breeding ground for the emergence of a specific social group, which increasingly manifested itself through the Enlightenment and beyond and which formed the psychological-social basis of the totalitarian state: the masses.

PART II

MASS FORMATION AND TOTALITARIANISM

CHAPTER 6

The Rise of the Masses

"Enlightenment is man's release from his self-incurred tutelage. Tutelage is man's inability to make use of his understanding without direction from another . . . 'Dare to think! Have the courage to make use of your own reason!' is therefore the motto of the Enlightenment."[1]

With these words in 1784, the great German Enlightenment philosopher Immanuel Kant summarized what he considered the essence of the Enlightenment tradition. A century and a half later, however, a horrifying phenomenon unfolded: The Enlightenment had led to the exact *opposite* of what Kant envisioned. "Science" had given rise to stories that were frankly absurd; people nonetheless went along with them in blind enthusiasm and fanaticism, with little ability for critical reflection, even up to the point of radical self-destruction.

In Germany, a race theory, propagated by a fanatical demagogue, propelled a large part of the population into a curious state of mind. People denounced relatives, friends, and colleagues who, in their opinion, were not unconditionally loyal to the German People and its leader; they accepted that fellow human beings with physical impairments be exterminated like vermin; they nodded in agreement when the Führer deemed the elimination of every German with heart and lung problems to be necessary in the long term; they agreed, overtly or covertly, with the industrialized annihilation of "inferior races."

In Russia, an equally "scientific" story led to the same fanatical ecstasy: The whole "historical-materialist process" would focus on the creation of a society without private property, in which "the proletariat" would have the power. This also required a fair bit of extermination. At first, this took place according to a certain "logic"; at a later stage, everybody randomly fell prey to it. Tens of millions of people were deported to the gulags, where the majority of people perished. Half of the members of the communist party were also eventually liquidated, usually without the slightest hint of sedition or treason. And the most astonishing thing of all was that most victims made no effort whatsoever to refute the mostly unfounded allegations. They even made unequivocal admissions of guilt and willingly went to the gallows.

The first half of the twentieth century saw the emergence of Nazism and Stalinism, a completely new form of government commonly referred to as *totalitarianism*. It is immediately distinguishable from democracies by its one-party structure and its disregard for basic democratic principles, such as the right to free speech and self-determination. However, the totalitarian state also radically differs from dictatorial forms of government, both in its structure (its internal organization) and in its dynamics (its process-oriented progression). In her monumental book, *The Origins of Totalitarianism*, Hannah Arendt situates the essence of this difference at a psychological level. While dictatorships are essentially based on instilling a fear of physical aggression—the population is struck by such a degree of fear that the dictator (or the dictatorial regime) is able to unilaterally impose a social contract—the totalitarian state is grounded in the social-psychological process of *mass formation*.[2]

We have to take this process into account in order to understand the astounding psychological characteristics of a totalitarian population: the willingness of the individuals to blindly sacrifice their personal interests in favor of the collective, radical intolerance of dissident voices, a paranoid informant mentality that allows government to penetrate the very heart of private life, the curious susceptibility to absurd pseudo-scientific indoctrination and propaganda, the blind following of a narrow logic that transcends all ethical boundaries (making totalitarianism incompatible with religion), the loss of all diversity and creativity (making totalitarianism the enemy of art and culture), and intrinsic self-destructiveness (which ensures that totalitarian systems invariably annihilate themselves in the end).

An analysis of the psychological process of totalitarianism is extremely relevant in the twenty-first century. There are several signs that a new kind of (technocratic) totalitarianism is on the rise: an exponential increase in the number of intrusive actions by security agencies (opening mail, searching IT systems, installing eavesdropping devices, tapping telephones);[3] the general advance of surveillance society;[4] the increasing pressure on the right to privacy (especially since 9/11);[5] the sharp increase in the last decade in citizens snitching on one another through government-organized channels;[6] the increasing censorship and suppression of alternative voices, in particular during the coronavirus crisis;[7] loss of support for basic democratic principles;[8] and the introduction of an experimental vaccination program and QR code as a condition for having access to public spaces, and so on. The moment Arendt had anticipated in 1951 seems to be rapidly approaching: the emergence of a new totalitarian system led, not by "ring leaders" like Stalin and Hitler, but by dull bureaucrats and technocrats.[9]

In the first five chapters of this book, I described how the emergence of the mechanistic worldview brought society into a specific psychological condition over the past centuries. Society was increasingly gripped by a fanatical, mechanistic ideology that degenerated into dogma and blind belief (chapter 1); experiences of meaninglessness and social isolation increased hand over fist (chapter 2); hopes were increasingly placed on a utopian, technological solution to the problems inherent in human existence (chapter 3); public space was increasingly dominated by a pseudoscientific discourse of numbers, data, and statistics that completely blurred the line between scientific facts and fiction (chapter 4); and epidemic fear and uncertainty made the population yearn for absolute authority (chapter 5). In the present chapter, I'll describe how, from here, the socially fragmented population suddenly reunites into a unit through the process of mass formation.

* * *

A crowd is a specific kind of group. Its distinguishing feature is a far-reaching "uniformization" of individuals. In the crowd, everyone becomes equal to everyone else, people think together, and they tend to identify with the same ideals. Gustave Le Bon—the French sociologist and psychologist who published one of the most important works on mass formation, *Psychologie des foules* in 1895—argued that the "individual soul" in the masses is completely taken over by the "group soul."[10] This uniformization is accompanied by an almost absolute loss of rational thinking and the ability for critical reflection, even among people who, under "normal circumstances," are extremely intelligent and capable of well-founded criticism.[11] It is also accompanied by a strong tendency to surrender to impulses that, under normal circumstances, would be considered radically unethical.

Mass formation is as old as humanity itself and has appeared in many different forms. Historical examples bear witness to this diversity: the short-lived mass formation during Saint Bartholomew's night as opposed to the long-term mass formation of the French Revolution; the totally unstructured mass of the dancing plague in Strasbourg as opposed to the organized masses we find in the army and church; the religious masses of the Crusades as opposed to the pseudoscientific masses of the twentieth and twenty-first centuries; the gigantic masses of Nazism and Stalinism; the small-scale mass formation that occurs time and again in trial juries, and so on.

This last example, the mass formation that occurs in trial juries, is interesting because its small scale allows for a detailed investigation. Time and again, it appears that juries, in their final verdict, are hardly (or not at all) influenced by the argumentative qualities of a plea. An attorney who delivers a perfectly fact-based and rationally structured message will have little effect. Juries are almost exclusively susceptible to frequent repetition of simple emotional messages and poignant visual images (including numbers presented in graphs).[12] Think of all the successful trial lawyers: This is exactly how they build their plea.

Masses have been around since time immemorial, but Le Bon noted that, beginning in the nineteenth century, they steadily gained momentum.[13] Where they used to have only a short-lived influence that was curtailed and suppressed by the leaders of society, they became steadily more persistent and influential in policy making during and following the Enlightenment. This prompted Le Bon in 1895 to warn that the masses could take hold of society, leading to the emergence of a new form of governance.[14] Le Bon was not devoid of prophetic gifts, as this is exactly what happened thirty years later with the rise of totalitarian states in the twentieth century.

* * *

Where did this intensification of mass formation come from? It was a logical consequence of the effects of rationalization and mechanization of the world, as discussed in the previous chapters. More and more people entered a condition of *social atomization* and as soon as their numbers exceed a critical limit, the process of mass formation begins. Mass formation is a complex and dynamic phenomenon that can be compared to the way convection patterns arise in water or gas when they are heated up. In the first instance, the heat in individual water molecules rises, but the molecules are not yet moving. Then small, moving patterns, which quickly disappear, emerge locally. Subsequently, increasingly larger and more durable patterns occur. Finally, we see patterns that permanently set most of the water into motion. In doing so, the convection patterns completely change the behavior of the individual water molecules, bringing them into a completely new state of motion. In the same way, mass formation brings individual people into a new psychological "state of motion." And just as with convection patterns in water and gas, these patterns are small and short-lived at first. At a later stage, they set larger and larger societal "volumes" in motion over a longer period of time. The medieval mass formations were mostly local and ephemeral in nature; the mass formations of the French Revolution were already larger in scale and lasted a little longer; those of Stalinism and Nazism were much more significant and a lot more enduring. With the coronavirus crisis, we have, for the first time in history, reached a point where the entire world population is in the grip of a mass formation over a prolonged period of time.

* * *

There are four conditions in particular that have to be present in a society for large-scale mass formation to occur. These four conditions were present prior to the rise of Nazism and Stalinism, and they are also present now. I've already mentioned them as consequences of the mechanistic ideology. I'll summarize them again below.

* * *

The first condition is generalized loneliness, social isolation, and lack of social bonds among the population. The Enlightenment is characterized by an emergence of this phenomenon, but today the scale has grown to such an extent that the US Surgeon General Vivek Murthy began referring to it as *the loneliness epidemic*, and Theresa May in Great Britain actually appointed a Minister of Loneliness.[15] Not insignificant to my argument, loneliness is strongly associated with the use of social media and communication technology.[16] (Remember the effect of digitalized conversations, which I covered in chapter 3.) The problem is greatest in industrialized countries, those that are most firmly in the grip of mechanistic ideology.[17] About 30 percent of people living in these countries report chronic experiences of loneliness and isolation, and this percentage is increasing every year. I refer to Arendt who argued that this first condition is the most important: "The chief characteristic of the mass man is not brutality and backwardness, but his isolation and lack of normal social relationships."[18]

This deterioration of social connectedness leads to the second condition: lack of meaning in life. This second condition follows mainly from the first. Man, as a social being par excellence, lives for the Other. Remove the bond with the Other and he will experience his life as meaningless (whether he sees the connection with his loneliness or not). For instance, I describe in chapter 2 how industrialization removed meaning from work, in part by breaking the connection between the person who produces something and the person for whom it is intended. Moreover, the mechanistic worldview also led to meaninglessness in a more direct way: the machine of the universe, as well as the person-machine who is stuck in it, runs without purpose or meaning. The material particles interact with each other according to the laws of mechanics, but they have no intention whatsoever. Viewing life through this lens, whether justified or not, renders life meaningless. The phenomenon of bullshit jobs (see chapter 2) is perhaps the best illustration of this: In the second decade of the twenty-first century, half of the people were of the opinion that their job was meaningless.[19] A 2013 Gallup World Poll found that only 13 percent of people worldwide were truly engaged in their job; 63 percent said they were not engaged (they "sleepwalk through their work and may put time into it, but are not passionate about their job"); and 24 percent are actively disengaged, meaning they actively demoralize and demotivate their colleagues.[20] This is very significant.

The third condition is the widespread presence of free-floating anxiety and psychological unease within a population. Free-floating anxiety is a form of anxiety that is not image-bound, in contrast to anxiety that is image-bound (for example, fear of thunder, snakes, war). Such anxiety is mentally difficult to manage and presents the constant risk of turning into panic, which is perhaps the most aversive psychological state for human beings. For that reason, a person in that state seeks to link their anxiety to an object. Free-floating anxiety can be traced back to the first two conditions. A person who has lost his bond with the Other and does not feel meaning typically experiences an indefinable unease and anxiety. This condition has been strongly present in the first decades of the twenty-first century. For example, the World Health Organization (WHO) reports that one in five people worldwide has been diagnosed with an anxiety disorder. These numbers are striking, all the more so because they are likely an underestimation. And the incidence of mental suffering in general, including the cases that go undiagnosed, is of course even higher. This can, amongst others, be concluded from the enormous consumption of psychotropic drugs. In a small country like Belgium, with eleven million inhabitants, no fewer than three hundred million (!) doses of antidepressants are taken every year.

The fourth condition, in turn, also follows from the first three: a lot of free-floating frustration and aggression. The link between social isolation and irritability is logical and has also been established empirically.[21] People perturbed by loneliness, lack of meaning, and indefinable anxiety and unease generally feel increasingly irritable, frustrated, and/or aggressive and look for objects to take these feelings out on. The sharp increase of racist and threatening language on social media during the last decade (tripling between 2015 and 2020, see chapter 5) is a striking example. What accelerates mass formation is not so much the frustration and aggression that are effectively vented, but the potential of *unvented* aggression present in the population—aggression that is *still looking for an object.*

* * *

How exactly do these conditions lead to mass formation? The catalyst for mass formation is a suggestion in the public sphere.[22] If, under the aforementioned circumstances, a suggestive story is spread through the mass media that indicates an object of anxiety—for example, the aristocracy under Stalinism, the Jews under Nazism, the virus, and, later, the anti-vaxxers during the coronavirus crisis—and at the same time offers a strategy to deal with that object of anxiety, there is a real chance that all the free-flowing anxiety will attach itself to that object and there will be broad social support for the implementation of the strategy to control that object of anxiety.

This process yields a psychological gain. Firstly, the anxiety that previously roamed through society as a tenebrous fog is now linked to a specific cause and can be mentally controlled via the strategy put forward in the story. Secondly, through a common struggle with "the enemy," the disintegrating society regains its coherence, energy, and rudimentary meaning. For this reason, the fight against the object of anxiety then becomes a mission, laden with pathos and group heroism (for example, the Belgian government's "team of 11 million" going to war against the coronavirus). Thirdly, in this fight all latent brewing frustration and aggression is taken out, especially on the group that refuses to go along with the story and the mass formation. This brings an enormous release and satisfaction to the masses, which they will not let go of easily.

Through this process, an individual pivots from a highly aversive and painful psychological state of social isolation to the maximum interconnectedness that exists among the masses. This creates a kind of intoxication, which is the actual impetus to go along with the mass-forming narrative. In the prolonged mass formations that led to the rise of the totalitarian states, this intoxication was often merely latent but sometimes fully manifests itself overtly. Think, for example, of a crowd that sings together or chants slogans in a football stadium, for instance. The voice of the individual dissolves into the overwhelming, vibrating group voice; the individual feels supported by the crowd and "inherits" its vibrating energy. It doesn't matter what song or lyrics are sung; what matters is that they are sung *together*. An equivalent to this exists on a cognitive level: What one thinks does not matter; what counts is that people think it together. In this way, the masses come to accept even the most absurd ideas as true, or at least to act as if they were true.

* * *

The essence of mass formation amounts to the following: A society saturated with individualism and rationalism suddenly tilts toward the radically opposite condition, toward radically irrational collectivism. To put it in Nietzschean-classical terms: Dionysus, in one fell swoop, overthrows the dictatorship of Apollo and seizes power in society. This is also immediately apparent from the following: In all major mass formations, the main argument for joining in is solidarity with the collective. And those who refuse to participate are typically accused of lacking solidarity and civic responsibility. This is one reason why the absurd elements in a story do not matter to the masses: *The masses believe in the story not because it's accurate but because it creates a new social bond.*

The strategy of dealing with the object of anxiety fully accomplishes the purpose of a *ritual.* The function of ritualistic behaviors is to create group cohesion. It is a symbolic behavior that aims to subject the individual to the group. As such, it must ideally have as little practical use as possible and require sacrifice on the part of the individual. Think of the ritual sacrifices of food, animals, and humans in primitive societies. That is exactly why the absurdity of the coronavirus measures does not encounter any resistance from part of the population. In a sense, the more absurd and demanding the measures are, the better they will fulfill the function of a ritual and the more enthusiastically a certain part of the population will go along with it. Think, for example, of the fact that some people wear a mask when driving, even if they are the only person in the car.

The ritual function of mass behavior is always present. The experts in the coronavirus crisis have also been more or less aware of this. At times, they let it slip that the measures actually have hardly any practical use. In March 2020, an expert virologist stated on Belgian national television that the lockdowns would barely reduce the number of deaths;[23] in August 2020, an expert virologist suggested that the face masks have a largely symbolic function;[24] in October 2020, the health minister of Belgium said the same about the closure of bars and restaurants (implying that countless people saw their livelihood ruined for symbolic reasons).[25] The message is clear: The individual must at all times show that he submits to the interest of the collective, by performing self-destructive, symbolic (ritualistic) behaviors.

Ultimately, the reasons individuals participate in mass formation are rarely, if ever, rational in nature. The justification of the strategy is promoted by experts with fancy titles, often on national television, making it seem like a given measure is generally accepted. For many people, this suffices as proof of correctness of the measures: "Surely the experts know what they're doing." "Surely, they can't *all* be wrong." "They obviously wouldn't say it if it weren't true?" And so on. In other words, the *argumentum ad populum* (appeal to popularity) and the *argumentum ad auctoritatum* (appeal to authority), known as logical fallacies since ancient times, are enough for most people to accept the story. In everything, you feel that the underlying motivation to go along with the story is the group formation and the group pressure, not the accuracy of the story.

* * *

The well-known conformity experiment by Solomon Asch demonstrates in a very convincing way the enormous impact of mass formation on individual judgment.[26] Asch conducted his experiment shortly after World War II. He did so in an effort to understand how the often-absurd theories of Nazism and Stalinism gained such a strong grip on the population and sought to gain insight into the psychological mystery of mass formation and totalitarianism.

Take a good look at figure 6.1. Which of the segments A, B, and C has the same length as line 1? That was the question Asch asked the participants of his conformity experiment. Each group of eight test subjects included seven of Asch's employees, all of whom had been instructed to answer "line segment B" without blinking an eye. The eighth participant, the only genuine test subject, usually gave the same answer as the seven persons before him. Only 25 percent consistently stated what even a blind person could see: Not line B but line C has the same length as line 1. After the experiment, some test subjects said that they did know the correct answer but did not dare go against the group. Even more interestingly, others admitted that they had started to doubt their own judgment under group pressure and eventually accepted the absurd group judgment as true.

These three groups are always present in mass formation. There is always a group that is in the grip of mass formation and "believes" the story (this group constitutes the totalitarized part of the population), a second group that does not really believe it but remains quiet and goes along with the masses (or at least, does not oppose them), and a third group that does not believe in the mass-forming story and also speaks or acts out against it. These three groups typically intersect with all pre-existing social groups. This is shown, time and again, in historical examples of large-scale mass formation.[27] And it also became apparent during the coronavirus crisis. At the beginning of the crisis, new societal "camps" emerged at lightning speed, crossing all the pre-existing camps—people either went along with the virus story or not. Left or right of the political spectrum, regardless of skin color

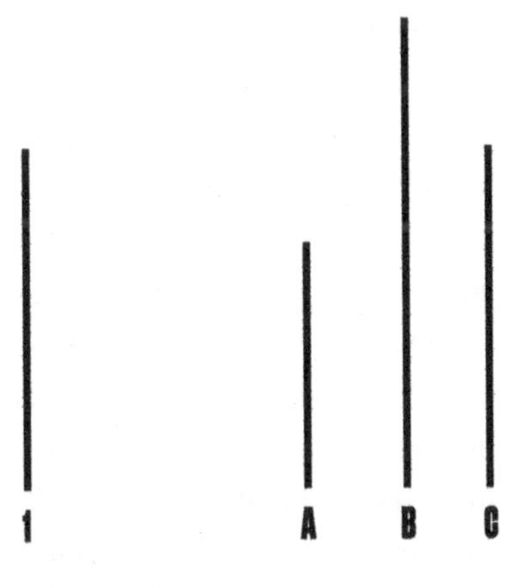

Figure 6.1.

and social status, profession and hobbies: All these boundaries blurred. The only thing that counted was what people thought of the virus.

Typically, these three groups are highly diverse, but for specific reasons this diversity is most visible in the group that protests loudly against the masses. In the mass itself, diversity disappears under the typical uniformizing effect of the masses (the masses make all individuals equal to everyone else) and the silent middle group does not stand out anyway, whereas the third, nonconformist group typically comes to life and all the individuals express themselves in their own specific way, which sharply highlights its diversity.

* * *

As Le Bon noted in 1895, the effect of mass formation is identical to hypnosis.[28] Both hypnosis and mass formation are largely caused by *a voice*, in the literal sense—through the physical, vibrating qualities of the voice. Totalitarian leaders are well aware of this, sometimes intuitively, sometimes consciously. Totalitarian systems have always been maintained primarily by systematic indoctrination and propaganda, injected into the population on a daily basis via mass media (without mass media, it is not possible to generate such long-lasting mass formation as that which gave rise to Stalinism and Nazism). This way, the population is literally kept on the vibrational frequency of the voice of totalitarian leaders.

On the one hand, the population is systematically exposed to the voice of the totalitarian leaders. On the other hand, every alternative voice is systematically eliminated. The first thing totalitarian leaders do is make sure their voices are the only ones left. To a certain extent, this is also what classical dictators do, but they limit the monopoly on the voice to the public sphere. They silence the political opposition. Totalitarian systems operate in a more thorough way. They censor alternative voices in the private sphere as well. On the one hand, this happens "spontaneously" due to a paranoid informant mentality that accompanies mass formation (which, in fact, is a result of a typical intolerance to alternative opinions, which we will discuss later). On the other hand, totalitarianism also expurgates the private sphere of alternative voices by inducing far-reaching social fragmentation and isolation. Totalitarian systems typically make it nearly impossible for people to gather in larger groups, and they strive to sever all social and family ties and replace them with the only allowable bond: the one between the individual and the totalitarian system (that is, the collective). In the Soviet Union, this process was implemented in a much more systematic way than in Nazi Germany; this is why the process of totalitarization in the Soviet Union persisted in a more far-reaching way.[29]

To return to the similarity between hypnosis and mass formation: In both cases, a suggestive statement or a suggestive story (conveyed by a voice) focuses attention on a very limited aspect of reality. Compare it to the circle of light emitted by a lamp, which is focused and makes everything outside of this circle disappear into darkness (see figure 6.2). In addition to the ritual function of the mass behaviors, this narrowing

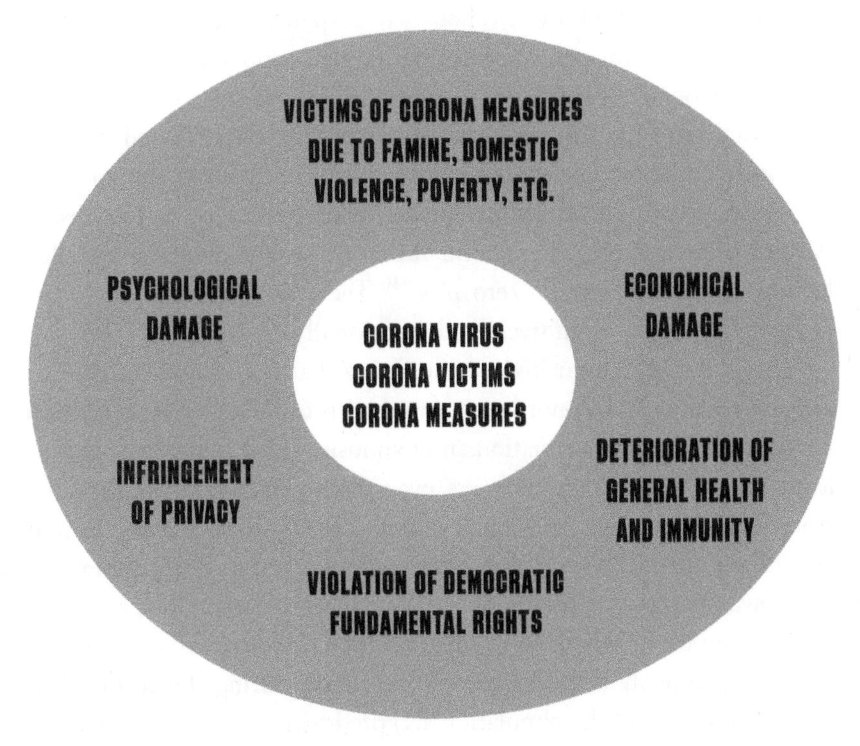

Figure 6.2.

of the field of attention is a factor that ensures the logic will extend to its absurd conclusions.

For example, in the coronavirus crisis, we've seen a narrowing of the field of attention in the following way: People who fall victim as a result of the measures—for example, deaths due to emotional and physical neglect during the lockdowns in residential care centers, non-COVID-19 patients whose treatment was postponed, victims of domestic violence, people affected by side effects of vaccinations, and so on—receive little attention compared to COVID-19 victims, or at least they are given remarkably little weight in decision-making. Furthermore, also very striking: The collateral damage of the victims may be mentioned occasionally, but it is rarely, if ever, presented in a numerical-visual way.

This is crucial because, as I described in chapter 4, what's expressed in numbers and graphs has the effect of being (wrongly) perceived as facts. As such, the psychological process of mass formation seems to ensure that mass media, almost intuitively, chooses to perpetuate the mass formation by using graphics for only the information that supports the story.

The narrowing of the field of attention also extends to the emotional field: Victims of the pandemic response measures have elicited remarkably little empathy. For these victims, there are no daily statistics, no case descriptions, no testimonials from relatives covered in the media. Also consider a virologist's statement that a boy who died at a so-called lockdown party deserved "zero pity."[30] These victims fall outside the circle of light, both cognitively and emotionally.

This emotional insensitivity to suffering that falls outside of the circle of attention should not be confused with ordinary egoism. Le Bon noted that both mass formation and hypnosis enable individuals to radically ignore their self-interest, yes, even their own pain.[31] The hypnotic story focuses attention on a small aspect of reality to such a degree that everything outside it, including one's own pain and to a broader extent, one's own interests, passes unnoticed. With a simple hypnotic procedure, patients can be anesthetized to such a degree that surgical incisions can be made painlessly (see chapter 10). Similarly, during the coronavirus crisis, a large part of the population accepted with remarkable ease measures that destroyed their enjoyment of life, freedom, and prosperity.

This was, by far, the most astonishing observation for the chroniclers of twentieth-century totalitarianism: The almost limitless tolerance for the enormous personal damage the population endured. For example, totalitarized Germans were devoutly grateful to Hitler for having a plan B in case their Great Mission would fail: death with grace—the gas chamber—for every German citizen.[32]

The phenomenon of mass formation not only has a profound impact on a cognitive and emotional level, but sometimes also on sensory perception. In some circumstances, collective hallucinations occur under the influence of mass formation, a phenomenon that challenges the understandings in modern psychology. A well-known historical example is the appearance of Saint Gregory on the city ramparts of Jerusalem, which was witnessed by a full army of crusaders.[33] Another example, from more recent times, is the raft of drowning persons observed in broad daylight by a whole crew of marines and described by each of them in the same way, in great detail. On closer inspection, it was no more than a few branches with seaweed on them.[34] The impact of mass formation on the mental functioning of humans is virtually unlimited. It has an impact on the individual's experience of reality to such an extent that it is justified to ask: For an individual in the grip of mass formation, is there still a reality beyond the one created by the masses?

* * *

We have to add one more important characteristic to the problematic psychological properties of mass formation: radical intolerance of other opinions and a strong tendency toward authoritarianism. To the masses, dissident voices appear 1) antisocial and devoid of solidarity, because they refuse to participate in the solidarity that the mass formation creates; 2) completely unfounded, as critical arguments are not assigned any cognitive or emotional weight within the narrow circle of attention of the masses; 3) extremely aversive because they threaten to break the intoxication, and in this way confront the masses again with the negative situation that preceded the mass formation (lack of social bond and meaning, indefinable fear and unease); 4) extremely frustrating because they threaten to remove the venting of latent aggression.

This radical intolerance ensures that the masses are convinced of their superior ethical and moral intentions and of the reprehensibility of everything and everyone who resists them: Whoever does not participate is a traitor of the collective. Snitching is therefore commonplace; the population itself is the main branch of the secret police.[35] Combined with the fourth factor, the opportunity mass formation offers to act out frustration and aggression without limit, this creates a well-known phenomenon: The masses are inclined to commit atrocities against those who resist them and typically execute them as if it were an ethical, sacred duty. Historical examples are the *Deus lo volt* (God wills it) and the *Got mit uns* (God with us) with which the Crusaders and the Nazis respectively committed their atrocities; the Bolshevik belief that the ultimate justice was being practiced by massacring the Romanovs and other alleged enemies of the proletariat; a butcher who, during the French Revolution, slit the throat of the defenseless (and innocent) director of the Bastille with a knife and demanded a medal for doing so;[36] the Septemberists of the French Revolution who conscientiously ensured that all citizens were able to closely watch the executions of clergy and noblemen.[37]

According to Le Bon, authoritarianism and intolerance are essential characteristics of mass formation.[38] We also see this characteristic of mass formation steadily on the rise in coronavirus society. As the crisis continues to unfold, the dominant discourse imposes itself in an increasingly authoritarian way and censors and represses alternative voices in an ever more radical way. Publications that don't fit with the dominant narrative are blocked on social media, even if they've been published in top medical journals such as *The Lancet*; doctors and researchers critical of coronavirus measures are fired by their institutes; in early 2021, the Belgian Order of Physicians issued a general rule that any doctor who casts doubt on the effectiveness and safety of the vaccine would be barred; from November 2021, one was no longer allowed to enter restaurants, bars, and a number of other places without a QR code, and so on. This is ultimately the difference between the solidarity of the masses and loving connectedness: The former is always at the expense of a particular group, the latter is not.

CHAPTER 7

The Leaders of the Masses

In the previous chapter, I described the phenomenon of mass formation—the psychological basis of totalitarianism—as a form of hypnosis. However, there is an important difference between mass formation and classical hypnosis. In classical hypnosis, only the field of consciousness of the hypnotized person is narrowed; the person who conveys the hypnotizing story (the hypnotist) is "awake." In mass formation, to the contrary, the person who conveys the story is usually in the grip of the story as well.[1] In fact, this person's field of attention is usually even more narrow than that of the masses. The reason is clear: The leader usually fanatically believes in the ideological basis of the narrative (not in the narrative itself!) that controls the masses.

With respect to the leaders, mass-formation gives rise to two opposing attitudes: Either one trusts the leaders blindly (and disappears into the mass), or one completely distrusts them and sees them as people who knowingly carry out an evil plan (i.e., conspirators). In a certain sense, both extreme perspectives are based on a similar misunderstanding: They fallaciously endow the leaders with a virtually absolute knowledge (and power); the first group does so in a positive sense, the second group in a negative sense.

Other misconceptions are that the leaders are primarily driven by money (i.e., "follow the money" and "cui bono") or sadistic pleasure (i.e., they have a psychopathic or perverted personality). Such statements are not really confirmed by historical research either. To give one example: The head of the Nazi party had a reluctant attitude toward illicit profits, and personalities with tendencies toward perversion and psychopathy were systematically excluded from recruitment.[2] As opposed to the "classical" criminal, who finds an intrinsic pleasure in violating social rules, in this case totalitarian criminality lies more in the uncritical and mindless adherence to a system of totalitarian social rules, even when this system becomes radically inhumane and transcends each and every ethical boundary. Hence Hannah Arendt's famous expression that totalitarianism was a true demonstration of the *banality of evil*: Totalitarianism is not about monstruous people—it is about normal people who stick to a morbid, dehumanizing way of thinking or "logic."[3]

In the initial phase of the totalitarization process, such a logic first takes hold of the population. The masses (or at least a large part of the population) become imbued with certain ideological convictions that, to them, are no longer distinguishable from reality. The emerging mass movements of Pan-Slavism and Pan-Germanism in Russia and Germany in the early twentieth century are good examples. Germans typically became convinced that, as a race, they were superior to others and that stigmatization and oppression of, among others, Poles and Jews could be justified by "the facts." We see something similar happening during the coronavirus crisis, where a certain segment of the population is becoming convinced that the facts justify the social discrimination of people who refuse to be vaccinated. The numbers show that they are spreading the virus, don't they?

These dynamics slowly give rise to the emergence of totalitarian parties and totalitarian leaders who gradually institutionalize this logic and impose it on society. And this typically happens in a fanatical, blind, and merciless way. Hitler believed that his strength came from his ability for "ice-cold reasoning," and Stalin believed that the secret of his success lay in his "merciless dialectics."[4] Races that were "unfit for life" and "dying classes" were, under the justification of this logic, expelled from society with surgical precision. For this reason, what characterizes the leaders of the masses is not greed or sadism, but their morbid ideological drive: Reality must and will be adjusted to the ideological fiction.

Such drive leads to a mental and emotional blindness, which can assume truly astonishing proportions. This is illustrated by the mind-boggling way Nazi leader Adolf Eichmann testified during his trial in Jerusalem about organizing the deportation of Jews to the concentration camps. During the trial, he was still so imbued with the conviction that he only intended to do the very best for everyone that he described with pride how he encouraged the Jews to participate in his "project." For example, in occupied European cities, he promoted the establishment of Jewish councils, consisting of Jews who occupied key social positions within their communities. Eichmann thought it was normal that the victims—who, within the Nazi doctrine, were regarded as people unfit for life—organize the practical arrangements of their own demise. In his trial, Eichmann described his attitude in the following way:

> *The Jewish council—who it consisted of, what their hierarchy was and how they organized the division of tasks—it was all left to them. We were, of course, in charge. But as I said, we did not treat them in a dictatorial way, we played it very carefully with the officials with whom we had to deal on a regular basis; therefore, our people intervened as little as possible, for the simple reason: if we had acted in an authoritarian way towards those top officials—like: you must—then that would not serve the purpose. Because if those involved do not want to cooperate, then the whole work suffers as a result; we tried everything to make the cooperation attractive.*[5]

The Nazis were indeed often convinced of their good intentions; a willingness to acknowledge this is a sign of maturity and essential to learning from history. But of course, this should in no way be interpreted as an argument to excuse their crimes. A human in the grip of mass formation may, in a sense, not know what he is doing, but that does not mean that he should be forgiven just like that. In a state of mass formation or hypnosis, people do still have the ability to make ethical choices. It is well known that, while under hypnosis, people may be made to do things they would be painfully ashamed of otherwise (undressing themselves, performing ridiculous dance moves) and be led to perform physical feats that they are normally incapable of (laying stiff as a plank between two chairs, for example), but they cannot be persuaded to cross ethical boundaries that they respect in an "awake" state.

The anonymity offered by the masses—the individual disappears into the crowd and feels unseen—is essentially just an excuse and a cover for letting one's own compulsions run wild. Whoever commits crimes in a crowd shows, above all, that under normal circumstances, he controls himself only for tactical, and not ethical, reasons. The explanation for the immorality of the masses does not mean that mass formation removes a normally present ethical awareness.[6] It means that it temporarily suspends *the concealment of its lack*. In this way, the masses reveal the real ethical dimensions of man.

Eichmann was not the only Nazi who believed in his remarkable ideological "benignness." The entire Nazi discourse on the extermination camps testified to this. They called death in the gas chambers the "death of grace" (i.e., the least painful solution for people they felt were better off dead than alive). The Führer had that same death in mind even for the entire German people if Germany were to lose the war: He promised on his word of honor that he had set aside a sufficient volume of gas in case of this scenario. Even at the Nuremberg Trials, the Nazi leaders continued to speak matter-of-factly of such death as a "medical act," a precision therapeutic intervention to render society "healthy."

Arendt notes that there was something even more remarkable than the appeal for cooperation that Eichmann addressed to the Jews: that he also *obtained* that cooperation. Arendt writes:

> *The Jewish Councils of Elders were informed by Eichmann or his men of how many Jews were needed to fill each train, and they made out the list of deportees. And the Jews registered, filled out innumerable forms, answered pages and pages of questionnaires regarding their property so that it could be seized the more easily. And then, right on time, they assembled at the collection points and boarded the trains. The few who tried to hide or to escape were rounded up by a special Jewish police force. As far as Eichmann could see, no one protested, no one refused to cooperate, that thanks to the "general cooperation" everything was right.* "Immerzu fahren hier die Leute zu ihrem eigenen Begräbnis." *They knew it. All of them.*[7]

The Jewish councils went along with Eichmann's "project," "until they, too, were deported, usually 'only' to Theresienstadt or Bergen-Belsen, if they happened to be from Central or Western Europe, but to Auschwitz if they were from an Eastern European community."[8] Sometimes there was heroic resistance and the gruesome manner in which it was crushed must have played a role in discouraging it. Think of 425 young Dutch Jews who, after fighting with a German security police detachment, were tortured for months on end in Buchenwald, to the point of death.[9] Still, the extent to which victims repeatedly complied with the plans of Nazi executioners should not be ignored from a psychological perspective; apparently many of them were also in the grip of mass formation.

The Jews were by no means exceptional in this regard. Many Germans remained loyal to Hitler even when his plans included purges among themselves; it was planned, for example, to exterminate Germans with heart and lung problems, and subsequently, those with all kinds of other deficiencies—plans that were not carried out because of the course of the war. Similarly in the Soviet Union, people passively awaited their turn to be picked up and taken to the gulags (read *The Gulag Archipelago* by Aleksandr Solzhenitsyn).[10] I myself listened in astonishment to a woman who had grown up in the Soviet Union and lost her father and uncle to the gulags, but shrugged her shoulders and said the system "had its pros and cons." Mass formation takes both victims and perpetrators in its grip.

The fact that totalitarian leaders are themselves caught in a form of hypnosis is illustrated by the psychological response to being removed from the crowd. When Nazi leaders were stationed for a prolonged time in countries that proved unsusceptible to mass formation, such as Denmark and Bulgaria, something predictable happened: They began to feel insecure about the cause they were serving, and the Nazi regime could no longer rely on them.[11] In other words, they woke up. This shows that the leaders are not only hypnotized by their ideology but also by the masses. The leader himself is entranced by the effects he produces in the crowd. Between the psychological condition of the masses and their leaders, there is a kind of *circular causality*: They hypnotize one another.

The fact that the totalitarian leader is himself under hypnosis and blind doesn't mean that he believes everything he tells the population. On the contrary. It is more accurate to put it this way: He blindly believes in the ideology he is trying to impose but not in the discourse he uses to promote it. He believes so fanatically in his ideology that he considers it justified to limitlessly manipulate, lie, and deceive in order to realize that ideology. Mankind (or part thereof) is on its way to the best of all worlds and therefore everything is permissible.

This can be seen in the way Nazism and Stalinism made use of numbers and statistics—including them profusely in their propaganda—consistent with the scientific allure of their story (and the story of every totalitarian system). Over time, the numbers showed a "radical contempt for the facts" to the extent that the facts were modified to make the numbers add up. In the Soviet Union, it was not uncommon to randomly pick "traitors" off the streets at the end of the week if a predetermined quota had not yet been reached.[12] In this way, the scientists who allowed themselves to be hypnotized by totalitarianism quickly turned into "charlatans."[13] They often ended up entrenched in a discourse that didn't even bother to disguise its deceitful and manipulative nature.[14]

Curiously, the masses are always willing to forgive their leaders. Undeniable evidence of manipulation and deceit is whitewashed with phrases like, "It may be mean, but it's smart" and "In the end, they do it for our good." Of this, Arendt writes:

> *The totalitarian mass leaders based their propaganda on the correct psychological assumption that, under such conditions, one could make people believe the most fantastic statements one day, and trust that if the next day they were given irrefutable proof of their falsehood, they would take refuge in cynicism; instead of deserting the leaders who had lied to them, they would protest that they had known all along that the statement was a lie and would admire the leaders for their superior tactical cleverness.*[15]

Just like the population (see chapter 6), the leaders are also capable of fanatical self-denial.[16] One of the most astonishing observations of the executions of convicted Communist Party leaders during the Moscow trials was the repentant acceptance they showed (also masterfully described by George Orwell in *Animal Farm*).[17] Although they were mostly completely innocent of the crimes they were charged with, they meekly accepted their sentence and pleaded guilty. What's more: They often diligently adduced evidence to prove their own guilt and cooperated in their own conviction, if only to ensure that their status as a party member would be preserved.[18] They perpetuated the hypnosis to the point of death. Waking up just before such a terminal moment would be very painful indeed.

This resulted in a perplexing dynamic in which party members continued to let friends, colleagues, and everyone else around them fall prey to the absurd brutality of the system (including fatal torture), until they themselves were devoured by the monster of totalitarianism. Of this Solzhenitsyn writes, "The majority of those in power, up to the very moment of their own judgment, were pitiless in arresting others, obediently destroyed their peers in accordance with those same instructions and handed over to retribution any friend or comrade in arms of yesterday."[19]

This shows us again: The essence of totalitarianism is not utilitarian or selfish in nature. Money and power only constitute intermediate ends. The ultimate goal is to realize an ideological fiction, and the totalitarian leader blindly sacrifices his own interests to achieve it.[20] This is what Le Bon refers to when he states that the leaders of the masses are themselves also hypnotized, especially by the ideology in which they fanatically believe.[21]

This anti-utilitarian nature is also reflected in the recklessness with which totalitarian regimes destroy their own economies and wreak economic havoc. For example, labor camps could be considered to be aimed at cheap labor and monetary yields, but nothing could be further from the truth.[22] They were organized in such a way that they were not profitable at all, barely even self-sufficient. Those labor camps were primarily experimental spaces, pilot projects for an ideal society, where an elite learns how to subject a population to its ideology.[23] Experimentation on humans is the prototypical activity of totalitarianism. It is the ultimate submission of reality to the pseudoscientific, ideological fiction.

This does not mean that the totalitarian leader is a typical idealist. He differs from an idealist in that he shows a radical, fanatical blindness but definitely also because of a remarkable lack of principle and aversion to laws. For example, he typically rules by decree on the basis of temporary rules that can be adjusted at his discretion.[24] The only law he really upholds is that there are no laws.[25] This is a risk we also run with the coronavirus pandemic, where emergency rules have superseded established laws and fundamental rights. In such an emergency, there is no right to protest, no need for a government to have its actions approved by parliament or Congress, no need to respect private property. Combined with the fact that medical tests with questionable validity have become an accepted basis on which to declare a pandemic emergency at virtually any time, the magnitude of risk to individuals and society is incalculable.

Each law forms an obstacle to the enforcement of totalitarianism's iron logic. "If we want to realize the ultimate goal of history—the reign of the proletariat, the creation of the super race, and so on—then we must eliminate all aristocrats and peasants, we have to exterminate all disabled people and Jews, and so on." But also, "if we want to prevent the ICUs from filling up, we have to go into lockdown and shut down the entire society, prohibit the elderly from seeing their grandchildren, no longer administer first aid in accidents, prevent women who have just given birth from holding their newborn babies, prohibit any further protests, prohibit people without a vaccine from traveling and working in healthcare, etc." If someone had presented such a line of reasoning prior to the coronavirus crisis, people would have wondered, with pity, about that person's mental health. Nowadays, it seems unshakable to many people. "You can't say A without saying B and C, and so on to the end of the deadly alphabet," Arendt said.[26] Once one has accepted the premise of the logic, everything else inevitably follows from there.[27] Every logical counterargument is systematically banned from the field of attention and rendered impotent, and step by step, all normal ethical boundaries are violated.

The frenetic urge of totalitarianism to impose a basic logic on society also manifests itself in the obsession with signs, sometimes used as a distinguishing feature for the elite (uniforms, medals, badges, etc.),[28] sometimes as a stigma for the "objectified enemies" of the regime, which is burned into the flesh, if deemed necessary (e.g., the tattooed numbers in Auschwitz; but also in the gulags, each group properly had its signs). With its sign system, totalitarianism tries to imprint its logic on reality, to permanently link it to the real world. Importantly, the assignment of signs and stigmas is usually the first step in the process of destruction.[29]

At this point, we are able to pinpoint the psychological essence of totalitarianism: an attempt to reduce the polysemy of human language to the monosemy of a sign system. As discussed in chapter 5, the difference between humans and animals lies primarily in the communication system. Animals use signs, which relate in an unambiguous and relatively invariable way to what they refer to (e.g., the silvery white belly of the stickleback female refers to sexual receptivity; it is consistent, across all individuals, contexts, eras, and locales), whereas people use symbols or words (signifiers) which, depending on the context in which they appear, may mean something completely different. This characteristic of human language introduces an endless richness and variety to human experience and culture, an endless possibility for the creation of new forms of expression and identities. But it also creates a fundamental uncertainty, which constitutes man's greatest anguish. No other living being is tormented by questions such as "Who am I?," "What do I want?," "What do I mean to the Other?"

Totalitarianism is the ultimate attempt to rid ourselves of this uncertainty by withdrawing into a (pseudo)scientific certainty and merciless logic, by trying to reduce symbols to signs, and by trying to annihilate all variety in cultural expression. Totalitarianism obliterates such diversity in every possible way.[30] The systematized and industrialized transportation, exploitation, and murder of population groups in labor and extermination camps are historical examples indelibly engraved in our minds.

The logic of a totalitarian system is in constant flux and typically becomes ever more absurd. The raison d'être of a totalitarian system consists of, among others, channeling anxiety, which is why it must constantly identify new objects of anxiety. When the system is no longer able to link anxiety to an object, it loses its raison d'être. Both Nazism and Stalinism were constantly restructuring themselves; the essence of the phenomenon of totalitarianism lies in its *dynamics*. The directives and decrees are constantly changing because it is imperative to create new responses to new threats. Think of the pigs in *Animal Farm*,[31] who wrote new rules on the wall overnight.

Also in recent decades, we have seen the emergence of many objects of anxiety in our society; they have appeared at an accelerating pace, leading to more and more restrictions on civil liberties: terrorism, climate change, coronavirus. Especially during the coronavirus crisis, we see the constant drive of new threats and need for new actions (the endless series of coronavirus variants, which necessitates the introduction of new measures). Moreover the whole trajectory on which the story developed was marked by curious changes: First the lockdowns were justified to "flatten the curve." The virus would spread anyway; it was just a matter of slowing the spread. Subsequently, we had to "crush the curve": It was suddenly no longer a matter of slowing the spread but of bringing infections to zero, something initially considered impossible. And when infections had virtually disappeared, measures were taken to *prevent* them (you could say we switched to "prevent the curve"). Over time, the rules changed at such a pace that no one seemed to know them anymore, and people accepted more and more passively that henceforward, they could be fined for anything and everything without an ounce of legal protection against such arbitrariness.

Throughout this whole process, the story shows itself immune to criticism, affirming itself to the point of absurdity. For instance, in a paradoxical way, the people who fall victim as a result of the measures (for example, due to isolation in residential care centers) are used as an argument *in favor* of the measures. These victims are carelessly added to mortality counts and are therefore used to *justify* the measures. In the same vein, the UN warned that famines resulting from lockdowns could kill millions of people.[32] We run the risk that these will also be wrongly counted among the COVID-19 victims and that the fear, and thus support for stricter measures, will increase exponentially. And the same problem is likely to occur with victims of the vaccination campaign. In this way, society could end up in a vicious circle: The stricter the measures, the more victims; the more victims, the stricter the measures.

The fact that this should be understood in terms of mass psychology rather than malicious, intentional deception (i.e., a conspiracy; see chapter 8) doesn't make it any less dangerous. On the contrary. The lack of critical reflection, the irrational allocation of empathy, and the willingness among part of the population to accept great personal loss are an extremely dangerous cocktail. The way in which unvaccinated people are denied access to parts of public spaces, which now even engenders support within the population for denying them access to grocery stores and hospitals, evokes the most unpleasant reminiscences and may indeed become the first step of an infernal cycle of dehumanization.

Don't underestimate where this could go in the future, and not just for the people in opposition. The idea put forward during the coronavirus crisis to place infected individuals in quarantine centers is still largely considered "unrealistic" and "disproportionate," but within a narrow virological line of reasoning, it could easily become the next logical step. Insofar as we're unable to think outside of the story, it requires only a heightened level of anxiety (or frustration and aggression) for this to be "necessary for public health." Combined with the manipulability of the COVID-19 tests and a feudal redistribution of power (mayors and governors are endowed with unprecedented power due to the impasse of national politics), we can see what appears on the horizon: random roundups, arbitrary isolation, and discretionary "treatment" of "infected" people. Societal systems that tend toward totalitarianism all lead to more or less the same phenomena, however different their stories might be in terms of substance.

The vicious circle in which mass formation and totalitarianism typically end up is, in a cynical way, also "reassuring": Mass formation and totalitarianism invariably destroy themselves by way of logical necessity.[33] They are intrinsically self-destructive. The underlying mechanism of self-destructiveness can be understood in this way: Mass formation feeds on anxiety and aggression; without the fear and the prospect of venting this aggression, the mass dynamics grind to a halt. The leaders realize that, if this happens, the masses will wake up and become aware of the damage they have suffered, whereupon they will turn against the leaders in a lethal fashion. Consequently, the leaders have no choice but to keep identifying new objects of anxiety and introducing new measures to destroy such objects. And the totalitarized part of the population follows them willingly, for reasons described in chapter 6 (see group 1 in mass formation): In this way their anxiety remains linked to an object, they are able to vent their frustration and destructiveness, and realize time and again a new social connection via new rituals of death. This is how the vicious, self-destructive cycle of totalitarianism (and mass formation) works.

The self-destructiveness of totalitarian systems typically reaches its peak the moment the system succeeds in gagging any dissenting voice and silencing the opposition. The Soviet Union reached this point around 1930 (when Stalin had acquired almost unlimited power and started his great purges), while Nazi Germany reached this point around 1935. Here as well, we see a radical difference with dictatorships, which almost always moderate their aggression from the moment they firmly hold the power. At that point, a dictator will, in fact, typically use his common sense: If I want to stay in power, I have to convince the population that it will be to their benefit. A totalitarian leader, on the contrary, is blinded by ideology and the accompanying mass formation, and for this reason, he lacks that common sense. When the moment of total power has arrived, he just continues to follow the madness of his logic to its limit. While dissenting voices are extremely aversive to individuals in the grip of mass formation, they are literally vital to him—a bitter drug he desperately tries to avoid but without which he is doomed. Without dissenting voices to break the massive resonance of the mass narrative, a totalitarian system lapses into radical self-destructiveness; the hypnosis becomes complete. The totalitarian state then becomes, as Arendt described, "a monster that devours its own children."[34]

Anyone who wants to understand how unpredictable and absurd such destructiveness can become can read Solzhenitsyn's account of the various waves of persecution and genocide under Stalin.[35] During this period, the regime constantly targeted new groups of the population, to be identified as "objective enemies"—people who had not committed any hostile act but were *deemed capable of doing so* by virtue of the group to which they belonged. Time and again, these new enemies were isolated and eliminated.[36] At first, it was possible to discern some logic in the great purges: They started by deporting the bourgeoisie, then the army officers who returned from abroad (they were too indoctrinated by capitalist logic), then anyone who had anything to do with religion (they were not convertible to communism), then all the people who might own gold (dentists, watchmakers, jewelers), then the peasants who were just a bit better off than other peasants, and at a later stage all peasants, tout court. These were all people who were too "petty-bourgeois" or possibly too affected by contact with capitalists. However, a little later—after all those groups had been deported or exterminated—the system still had to discharge its destructive instinct, and random "criminal" population groups became the object of destruction.[37] The tempestuous, destructive dynamics of totalitarian systems also occurred in Nazi Germany but did not develop all the way to their ominous end.[38] After Hitler deported the gypsies and the Jews to the concentration camps, he aimed to target not only the Ukrainians and the Poles, but also all Germans with heart and lung problems. The war ultimately ensured that these plans could never be carried out.

There are many reasons to assume that totalitarianism starts from megalomaniac albeit "good" intentions. It aspires to no less than a total transformation of society into an ideological ideal (for example, the racially pure society of Nazism or the rule of the proletariat under Stalinism). However the creation of the paradise typically ends in an inferno. The history of Stalinism illustrates this in the most poignant way. The Bolsheviks started out with a determination to remedy the abuses of tsarist Russia. Under the tsars, about seventeen death sentences were carried out each year. The communist revolutionaries thought that outrageous. They screamed bloody murder: The death penalty should be abolished. However, the contract contained a small footnote: In the beginning, there still would be executions if it was necessary to install communism itself as a system. In the first months after the Russian Revolution of 1917, there were 540 executions per year; after a few years, this increased to 12,000 per year; and between 1937 and 1938 more than 600,000 executions were carried out per year.[39]

Even more astounding than the numbers of victims was the arbitrary way in which people were sentenced to death. Each city and region was given weekly and monthly quotas that stipulated how many "traitors" had to be arrested. If, at the end of such a period, the local mandate holders observed that the target number had not yet been reached, they took to the streets and arrested people at random:

> *This submissiveness was also due to ignorance of the mechanics of epidemic arrests. By and large, the Organs had no profound reasons for their choice of whom to arrest and whom not to arrest. They merely had over-all assignments, quotas for a specific number of judgments. These quotas might be filled on an orderly basis or whole arbitrarily.*[40]

The revolutionaries aimed to not only abolish the death penalty, they would also end all forms of slavery. But that didn't turn out as expected either. Solzhenitsyn presents a baffling comparison between the living conditions of "the proletariat" under the tsars and under Stalin. He describes how serfs under the tsars were only allowed to work a maximum of seven hours a day in the winter and twelve hours a day in the summer. When assigning orders and work, workers' physical limits were always taken into account. Moreover, the labor camps themselves were, all in all, tolerable. Fyodor Dostoevsky described them as so comfortable that the nobles began to fear they would eventually fail to instill fear. Under Stalinism, there was indeed a profound change in the fate of prisoners, unfortunately not for the better. A poignant point of comparison: Under the tsars, prisoners were required to mine 118 pounds of ore a day; under the communists this became 28,800 pounds![41]

Another good intention of the Bolsheviks was to improve the lot of peasants. But along the way, they changed their minds. In their attachment to their land and their animals, the kulaks proved to be too "petit bourgeois," and therefore unfit to passionately love the communist Leviathan.[42] The communists decided by decree that the peasantry as a class should be exterminated. They hastily introduced a deportation policy, which, in many ways has no equal in history. Peasants were driven by tens of millions into so-called "special settlements" where they often died down to the last person due to unequivocally inhumane conditions.[43] So, once again, they were reduced to serfs, serfs whose conditions were, in almost every way, a lot worse than they were under the tsars.

Le Bon famously stated that "Crowds are only powerful for destruction."[44] Devoted to solidarity, they aim for the greater good in the belief that it will lead to an ideological paradise. The outcome, however, is invariably the same: an infernal abyss. Crowds and their rulers are blindly dragged into a maelstrom of destruction, until they are confronted with the ultimate consequence of the rationale that has monopolized their mind: the mechanistic logic of a dead, soulless universe. As I will elaborate on in chapter 8, the real masters of the predicament are not the leaders of totalitarian systems but the stories and their underlying ideology; these ideologies take possession of everyone and belong to no one; everyone plays a part, nobody knows the full script.

CHAPTER 8

Conspiracy and Ideology

If only there were evil people somewhere insidiously committing evil deeds, it were necessary only to separate them from the rest of us and destroy them. But the line dividing good and evil cuts through the heart of every human being, and who is willing to destroy a piece of his own heart?

ALEXANDRE SOLZHENITSYN[1]

Try out the following: Put three dots far apart on a sheet of paper. Randomly put a fourth dot on the sheet, anywhere you like. Then take a ruler, measure the distance between this fourth dot and any of the three other dots and divide it by two; put a new dot there. Measure the distance between this new dot with any of the three initial dots (randomly indicated) and divide the distance again by two, put a new dot there.

Repeat this process a few hundred times and you will witness an astonishing phenomenon. You will see that, from the nebula of points, a Sierpinski triangle will arise—a fractal pattern that, from its overall composition to its tiniest detail, shows an identical pattern, in this case a triangle with an inscribed triangle (see figure 8.1).

You can easily carry out this process with ten, a hundred, or even more people, who each take turns adding a point to the sheet of paper, blindly

Figure 8.1.

following the rules stipulated above, without knowing what the purpose of their actions is. You will create this pattern together by all individually applying the same simple rule over and over again. This is relevant to what I will discuss in this chapter: Upon seeing how a Sierpinski triangle arises on the sheet of paper, a naive viewer would inevitably be under the impression that the people making the points have detailed prior knowledge of this pattern and are working together in a planned and coordinated way. But the reality is different: Nobody needs to know or have ever even seen this pattern. It is enough that all people independently follow the same simple rules as they place their points. Keep this Sierpinski triangle in mind as you read this chapter, it will resonate here and there.

* * *

Are the leaders of the masses conspirators? Are mass formation and totalitarianism set in motion by a grand sophisticated scheme coordinated by a few people behind the scenes? This is a legitimate question. Hannah Arendt, for example, frequently pondered this in her work on totalitarianism.

One thing is certain: Throughout history, the leaders of the masses have often been *perceived* as conspirators. As the masses grew in strength and intensity throughout the nineteenth and twentieth centuries, conspiracy theories also emerged. These conspiracy theories were typically used to explain complex social processes and mass formations. The mother of them all is *Protocols of the Elders of Zion*, whose popularity, according to Henri Rollin, was second only to the Bible in the early twentieth century.[2] It proclaimed that there was some kind of secret Jewish world government that controlled and ruled over all national governments.

Despite its massive popularity, "the Protocols" were a fabrication. Their fictitious origin is beyond dispute. They are based on a text published by French lawyer Maurice Joly in 1864 under the title *Dialogue in Hell between Machiavelli and Montesquieu*, a kind of pamphlet in which the author denounced Napoleon III's hunger for power.[3] The text was edited and distorted by the Russian secret service Okhrana in the late 1800s with the intention of fueling anti-Semitism in Russia. The Okhrana retained about half of the original text, added a few paragraphs left and right, and consistently replaced *France* with *world* and *Napoléon III* with *Jews*. In this way, they manufactured a text in which Theodor Herzl, founder of Zionism, was the head of a Jewish conspiracy that aspired to world domination. The forged pamphlet was published in 1905, at which point Russian conservatives and the Russian Orthodox eagerly adopted it to justify their anti-Semitic agenda. From there, it made its way into Germany during the first half of the twentieth century and to the Middle East, where it remains extremely popular to this day.

The tendency to reduce large-scale mass formation to the machination of an evil elite, however, dates back to earlier times, at least from the beginning of the Enlightenment. For example, in 1813 Chevalier de Malet described theories asserting that the heroes of the French

Revolution were actually secret agents of masonic lodges, who in turn belonged to a wider "revolutionary sect" whose aim it was to manipulate the public rulers like pawns from behind the scenes.[4] This theory was, in turn, based on the *Monita Secreta*,[5] an even older booklet describing a Jesuit conspiracy in an attempt to instigate a hate campaign against the establishment. *Monita Secreta* was first published in 1612 and sold at book markets all over Europe until the end of the twentieth century.

* * *

The above theories are, in fact, full-fledged conspiracy theories. But nowadays the term *conspiracy theory* is bandied about, even when it concerns theories that don't make any mention of conspiracy at all. For this reason, it is good to first pursue some conceptual rigor and to define the term. According to Wikipedia, a conspiracy is: "A secret plan or agreement between persons [. . .] for an unlawful or harmful purposes, [. . .] while keeping their agreement secret from the public or from other people affected by it."[6] This definition shows that at least three core characteristics must be present for an activity to be classified as a conspiracy: 1) There has to be a conscious, intentional and *planned* endeavor. 2) This endeavor has to be hidden or secret. 3) The endeavor has to be aimed at inflicting harm (i.e., there must be some malice toward someone involved).

In current usage, however, the term denotes a wide range of theories. It is sometimes used accurately to refer to theories about global shadow governments (such as the Illuminati, or the Cabal) that would steer world history in its entirety, or even more exotically, about elites of extraterrestrial origin, more reptilian than human, who have the world in their grip (see for example, QAnon discourse). But the term is also currently used—incorrectly—to deride critiques of power structures at the levels of banking, politics, industry, economics, and media.

The term therefore has become a stigma, a discursive means by which the dominant discourse protects itself from critical reflection. Likewise, the term *conspiracy* is rarely, if ever, used to refer to theories that are in line with the dominant story and yet are actual conspiracy theories. For example, consider claims that Russia is trying to steer US elections, that the Chinese government is behind cyberattacks, that Steve Bannon is secretly circulating reports that the virus originated in a lab in Wuhan, that Russia funds all kinds of anarchist newspapers in the West, and so on. Whether these assertions are accurate or not, they are, in essence, conspiracy theories. The only reason they are not stigmatized as such is that they belong to the dominant social discourse as it is constructed every day through the mainstream media.

* * *

That said, we return to the question: Should we consider mass formation the result of a conspiracy? In the crowd, the individual soul is replaced by a common group soul, noted Gustave Le Bon.[7] The crowd acts in a coordinated way and repeats the same slogans. It engages thoughts and expressions that spread through its ranks at lightning speed (Le Bon described the "contagiousness" of thoughts in a crowd).[8] Every segment of society participates in that *pensée unique*—politicians, academics, the press, experts of all kinds, judges, and police officers. In this way, the masses give the impression of a highly organized phenomenon. Those who, for one reason or another, are not sensitive to the mass formation and who observe this social phenomenon "from the outside" tend to think this must be the result of a large-scale, conscious, and planned coordination.

In chapter 6, I explained that mass formation is largely the result of individuals being gripped by a common narrative that unites them in a heroic battle against an object of anxiety. Exactly how much this line of reason explains about the phenomenon of mass formation remains to be seen. For instance, there seems to be a real physical resonance among individuals who form a mass that cannot be explained solely on the basis of sharing the same narrative. The phenomenon has direct similarities with the way complex, dynamic systems organize themselves in nature. A well-known example is the way starlings swarm. At dusk, the starlings fly toward each other from all directions and begin to move together in a harmonic pattern, so perfect that Nobel Prize–winner

Nikolaas Tinbergen called the flock a "super individual," a kind of overarching entity in which all the individuals are connected to one another like cells of the same body.[9] They sense each other perfectly, without any observable form of communication directing their behavior.

The way in which individuals in a crowd establish connection with one another is similar. This is particularly visible when a crowd physically gathers. Elias Canetti describes it in the following way:

> *The crowd, suddenly there where there was nothing before, is a mysterious and universal phenomenon. A few people may have been standing together—five, ten or twelve, not more. Nothing has been announced, nothing is expected. Suddenly everything is swarming with people and more come streaming from all sides as though streets had only one direction. Most of them do not know what has happened and, if questioned, have no answer; but they hurry to be there where most other people are. There is a determination in their movement that is clearly different from the expression of ordinary curiosity. It seems as if the movement of one transmits itself to the others. But that is not all; they also have one goal, which is there before they can find words for it. The goal is the most intense darkness where the most people are gathered.*[10]

This means that the crowd is not only united by the same thoughts, beliefs, and behaviors. It also seems to form a kind of physical unity, which contributes to the overwhelming impression that it is the product of an immense, planned scheme.

* * *

It is not only coordination in the mental and physical movements of the crowd that make it come across as the product of a conspiracy. Its threatening nature also contributes to that impression. The crowd typically tries to impose its will on society; it seeks *control* over society. This has always been the case, but this may have become more obvious over time as crowds have taken on a more durable character and started to exert a constant influence on the fabric of society. The modern crowd is always pushing in the same direction: the hyper-controlled society. With each new object of anxiety—terrorism, climate problems, viruses—the call for greater technological control rises up from its belly. And such control can swing sharply and unexpectedly. After the 2016 terrorist attacks in Brussels, hundreds of cameras were installed in Antwerp's Jewish quarter to ensure better protection against terrorists. During the coronavirus crisis, those same cameras were used to monitor whether Jews were visiting the synagogue.[11] Things can turn in a strange direction.

The coronavirus pass (and QR code) is also part of this trend toward ever more control. The plan to replace this pass in the long (or short) term with a more sophisticated system, more efficient and difficult to falsify, rests easily within the logic of the mechanistic ideology. In 2021, a Belgian minister had already argued that an electronic bracelet would actually be better (why not an ankle bracelet?). The part of the population that is in the grip of the mechanistic ideology will certainly go along with it, and the current state of technology undoubtedly offers the prospect of even more efficient "solutions" to this problem. At the end of this process, we will be moving in the direction of a society as described by, amongst others, the Israeli historian Yuval Noah Harari, in which subcutaneous sensors constantly monitor the state of our blood and will not only be able to detect diseases at an early stage but will also know our state of mind, whether we are feeling sad or happy, angry or calm.[12]

People who are not in the grip of mass formation initially find themselves in an extremely diffuse situation that they do not understand—the phenomenon of mass formation appears absurd and bewildering to those who are not in its grip—and they feel threatened by its controlling appearance and its typical intolerance toward those who refuse to partake (see chapter 6). In this state, the confused spectator typically develops an intense need for a simple frame of reference, which allows him to mentally master the complexity, and in which to place and control the anxiety and other intense emotions that arise. An interpretation in terms of a conspiracy meets that need. It reduces the enormous complexity of the phenomenon to a simple frame of reference: All anxiety is linked to one object (a group of people who intentionally deceives, the supposed "elite") and thereby becomes mentally manageable. All blame can be placed outside oneself, with the Other and, subsequently, all the frustration and anger can also be directed at that singular object. For this reason, fanatical conspiracy thinking testifies to the almost irresistible tendency of human beings to find someone who can be held responsible in the face of adversity and can thereby be made the object of aggression. This can probably be seen as one manifestation of a more general, psychological rule: The more anger people feel, the more intentional malice they perceive.

As such, in a certain sense, conspiracy thinking—the thinking that reduces all world events to one big conspiracy—fulfills the same function as mass formation. As with mass formation, conspiracy theorizing fills humans with a kind of enthusiasm. The anxiety, anger, and discontent that are now associated with a few simple mental images transform a strongly negative state into a (symptomatic) positive one. Everything is now explainable by means of a simple frame of reference; the world is no longer absurd but logical; one knows where the enemy is and has a point to direct his frustration and anger on; you can absolve yourself of responsibility and forego the need to question your own self. This is how conspiracy thinking acquires enormous psychological importance. Due to the multiplicity of effects attached to these mental images, the images draw all mental energy in like a mental magnet and eventually impose themselves as explanations for almost everything that happens.

For these reasons, thinking in terms of conspiracies becomes tempting. That's why the conspiracy logic has a tendency to drift further and further off course, eventually ending up in the realm of the absurd, even among highly intelligent rational people. Ultimately, there is such fundamental distrust that many people assume that whatever "the mainstream" considers right must certainly be wrong: For example, if the mainstream story says the Earth is round, it must be flat. Conspiracy thinking also leads invariably to the dehumanization of a certain group (in fact, *dehumanization* sometimes has to be taken literally: The elite consists of reptiles or aliens). The elite is pure evil, they intentionally make us sick through toxic substances in our food and the environment, and are responsible for brainwashing children through education for ages, and so on. In this way of thinking, the knowledge and power of the elite are easily overestimated. The elite do not struggle with the lack of knowledge that characterizes human beings, they do not doubt or hesitate, they don't face unexpected hurdles, they do not miscalculate. They are able to manipulate all world events. Conspiracy thinking inflates the sizableness of the perceived enemy into infinity so that in the end one can only feel powerless compared to such a giant. In this way, conspiracy thinking also embodies an aspect of self-destruction.

* * *

Thinking in terms of conspiracies often arises from the appeal of those psychological "benefits" more than the facts (which, of course, applies to many forms of thinking). The internal logic is often strong, but the theories often fail to meet the facts. For example, if you get to know the people who are the subject of a conspiracy theory more closely, the theory usually spontaneously loses all persuasiveness. For example, during the coronavirus crisis, many people started to believe that the experts intentionally misled the population because they systematically made blatant statistical and other errors. Experts cannot be that stupid, can they? However, if you get to know the experts, you often immediately sense that you can't squeeze their mistakes into the simple reference frame of lucid manipulation. In July 2021, just before the summer holidays, I met with a few statisticians involved in the modeling that mapped the course of the infection counts. One of them reported his concern: The number of infections was rising again. I immediately replied: "A lot of people go on holiday during this period, and they are all being tested. Have you accounted for the influence of the higher number of tests performed?" He looked at his colleagues with despair and objected: "No, but nobody does that when conducting estimates of the number of infections," and "The predictions of the number of infections based on those models do follow the number of hospital admissions, do they not?" and "We saw last year what happened in the fall when we didn't follow those models," and so on. The fact that all his arguments were textbook examples of false arguments (*argumentum ad populum, argumentum ad auctoritatum, false consensus*) completely escaped this intelligent man. Nothing could convince him that more tests naturally lead to more positive tests. Remember Asch's experiment in chapter 6? Mass formation blinds both intelligent and less intelligent people to the same extent. People really don't have to be part of a conspiracy to systematically make the most foolish mistakes.

Furthermore, the one-sidedness with which the mainstream media reported about the coronavirus crisis seemed to indicate at first that there was an intentional and planned manipulation of the reporting. Why do we hear hardly any "dissident" voices? How can one repeat the same misinformation over and over again? And yet, I know several "corona-critical" journalists who told me that there was no systematic, planned steering of the reporting. There was sometimes implicit pressure, that's true. For example, some politicians suggested that it was not the right time to sow confusion by broadcasting all kinds of criticism with respect to the national policy. In a sense, that was undemocratic influence over the press—journalists knew that politicians would give them fewer scoops if they allowed too many critical voices to be heard—but that is still more accurately described as self-censorship rather than censorship.

I had the same impression in my own contact with politicians: They are generally people who have doubts, who wonder to what extent they can afford to deviate from the measures taken by other countries, who are afraid of being held accountable for coronavirus victims if they introduced more lenient measures, who respond to the demand of the masses to act decisively against dissidents. And there are indeed also a few who see their chance to impose their ideology on society. However, most politicians merely follow the story obediently, and to do so, they don't have to gather at "secret" meetings.

Incidentally, I also had the privilege of being the subject of a few conspiracy theories myself. Like many people who speak out critically in one way or another, I was accused of being so-called controlled opposition (i.e., cooperating covertly with the coronavirus policy). My sole intention, it seemed, was to keep the opposition calm and quiet with my psychological theories. Some went further and thought I was a satanist.

In interviews, I had made a number of more or less correct predictions about the course of the coronavirus crisis, for example that the measures would not be lifted after the vaccine rollout. To some conspiracists, it was clear: I had been informed in advance about the plan. And to confirm the devil worshippers, I had also announced the evil that was about to happen beforehand. To this day, I am unaware of my membership in any sinister society and I believe that my "predictions" were made on simple grounds. In the psycho-logic of the coronavirus story, I found nothing that could prevent the continuation of the measures after the vaccine rollout. The fear was already present prior to the coronavirus crisis and it would not go away with vaccination, regardless of whether the vaccine was effective or not. I think I am somewhat entitled to have a word in this matter but nevertheless understand that an explanation in terms of satanism is more appealing to some people.

It's also worth mentioning that people who identified with the dominant narrative also sometimes saw me as a conspirator. They were of the opinion that I didn't believe my own theory of mass formation, it was just shrewd manipulation to detract from social support for the measures; I was only aspiring to obtain a position in some right-wing political party. I can only say: I myself would be very surprised to find my name on any ballot in the next election.

* * *

Is there not any steering and manipulation at all then? The answer is a resounding yes, there most certainly is all kinds of manipulation. And with the means available to today's mass media, the possibilities are simply phenomenal. Such steering, however, is primarily not a steering by individuals; the most fundamental steering is impersonal in nature. The steering is first and foremost driven by an ideology—a way of thinking. Ideologies organize and structure society progressively and organically. As we have described in detail in the previous chapters, the dominant ideology is mechanistic in nature. This ideology typically derives its appeal from the utopian vision of an artificial paradise (see chapter 3). The world and man are a machine and they can be comprehended and manipulated as such. The hitches in the machine that cause suffering can be mechanically "repaired." Yes, even death can be eliminated in the long run. Moreover, all this can be done without man having to reflect on his role in his own misfortune, without questioning himself as a moral and ethical being. This ideology makes life easy in the short term; the price for convenience will be paid in arrears (see chapter 5).

It is at this fundamental level that we have to situate the "secret" forces that direct individuals in the same direction and ultimately organize society as a whole. As with drawing the Sierpinski triangle, if everyone follows the same rules, it results in strictly regular patterns emerging in society. Like iron filings scattered in the force field of a magnet, individuals arrange themselves in a perfect pattern under the influence of these forces. Man has always fallen prey to the aforementioned "temptations"—the illusion of rational understanding and control, the resistance to questioning oneself critically as a human, the pursuit of short-term convenience, and so on. Within the religious discourse, these temptations were considered dangerous, but that changed with the rise of mechanistic thinking. From then on, they became anchored in the dominant narrative, which also became their justification. Leaders and followers were captivated by the limitless possibilities the human mind seemed to offer. The whole evolution toward a hyper-controlled technological society—the surveillance society—is simply unavoidable as long as the human mind remains trapped in that logic and is (to a large extent unconsciously) controlled by those attractors. It is this ideology that redesigned society, created new institutions, and selected new authority figures. The transition from a democracy to a totalitarian technocracy, in which the coronavirus crisis was a Great Leap forward, actually formed part of the logic of the mechanistic ideology from the very beginning. In a mechanistic universe, it is inevitably the technical expert who has the last word, based on his superior mechanistic knowledge.

Based on this ideology, institutions were created that make plans about what future society should look like and how the ideal future society should respond to crisis situations. Operation Lockstep from the Rockefeller Foundation,[13] Event 201 of the Bill and Melinda Gates Foundation (in collaboration with Johns Hopkins University and the

Rockefeller Foundation),[14] and *COVID-19: The Great Reset* by Klaus Schwab[15] are examples of such endeavors. For many people, these events and publications are the ultimate proof that the social developments we're experiencing are planned and the product of a conspiracy. Since long before the outbreak, these "plans" described how society would go into lockdown as the result of a pandemic, that a biopassport would be introduced, that people would be tracked and traced with subcutaneous sensors, and so on.

If we keep in mind the definition of a conspiracy—a secret, planned, intentional and malicious scheme—we immediately notice two things: It's not much of a secret since all the aforementioned "plans" are openly available on the internet. And whether those plans guide the discourse and action of experts through targeted instructions is at least questionable. The experts' communication is full of contradictions and inconsistencies, retractions and corrections, clumsy wording and transparent errors. This is nothing like a streamlined execution of a pre-established plan. If these are conspirators, they are the lousiest ones ever. Obviously, psychological warfare may also make use of confusion and confusing messages, but that does not explain experts trying to correct their mistakes of the day before, with visible unease and discomfort.

The only consistency within the experts' discourse is that the decisions always move toward a more technologically and biomedically controlled society, in other words, toward the realization of the mechanistic ideology. As such, we see exactly the same problems in the coronavirus crisis as those revealed by the replication crisis in academic research: a maze of errors, sloppiness, and forced conclusions, in which researchers unconsciously confirm their ideological principles (the so-called *allegiance effect*, see chapter 4).

In the whole process of exercising power—i.e., shaping the world to the ideological beliefs—there usually is little need to make secret plans and agreements. As Noam Chomsky put it, if you have to tell someone what to do, you've chosen the wrong person.[16] In other words: The dominant ideology selects who ends up in key positions. Someone who does not share the ideology is usually less successful in society, apart from a few exceptions. Consequently, all people in positions of power automatically follow the same rules in their thinking and in their behavior and are under the influence of the same "attractors" (to use a term from complex dynamical systems theory). Furthermore, they all succumb to the same logical fallacies and the same absurd behavior because they all, independently of each other, or at least without having to gather in secret meetings, follow the same distorted logic. Compare it to computers running on the same, wrong software: Their "behavior" and their "thinking" will all deviate in the same direction, without "communicating" with one another. This is exactly what the Sierpinski triangle shows us: Mind-blowingly precise and regular patterns can arise because individuals independently follow the same simple rules of behavior by being attracted to the same set of attractors. The ultimate master is the ideology, not the elite.

Those plans and visions for the future are not so much "forced" on the population. In many ways, the leaders of the masses—the so-called elite—give the people what they want. When fearful, the population wants a more controlled society: The lockdowns were, for many, a liberation from the unbearable and meaningless routine of working life, the fragmented society was in need of a common enemy, and so on. The "plans" do not precede the developments, as a conspiracy logic likes to suggest. They rather *follow* them. Those who guide the masses are not real "leaders" in the sense that they do determine where the masses will go. Instead they sense what people crave and they adjust their plans in that direction, in an opportunistic way. They wallow in the narcissism of one who controls and directs the chain of events, but they are more like a child sitting on the bow of a ship and turning a toy steering wheel every time the tanker changes direction. Or we can think of King Cnut, who stood before the sea at low tide, ordered the waves to retreat, and narcissistically beamed with pride because it actually happened. It goes even so far that some of those institutions have even adapted previously released films, thereby suggesting that they can predict the future. (For example, the Digi-kosmos film was adapted in such a way that it seemed to predict the course of the coronavirus crisis exactly as it happened.[17]) Ironically, conspiracy thinking confirms the leaders' narcissism by taking them seriously and believing that they are truly steering the ship, or causing the waves to recede.

There are countless other examples that seem to point in the direction of a plan being implemented, such as the fact that the definition of *pandemic* was adjusted shortly before the coronavirus crisis; that the definition of *herd immunity* was changed during the crisis, implying that only vaccines can achieve it; that the counting method for COVID-19 deaths was adjusted by the WHO so it was higher than the number of flu deaths; that the registration methodology of vaccine side effects could not but lead to serious underestimation (for example, by labeling the side effects that become apparent during the first fortnight after vaccination as not vaccine related); that all key political positions when the crisis started were held by politicians who were pro-technocracy (referred to as the World Economic Forum's Young Global Leaders); and so on.

These, too, are examples of how an ideology gets a grip on society rather than evidence of the execution of a conspiracy. For instance: Similar things happen during almost all major reorganizations in large companies and government institutions. Indeed, anyone who would like to reorganize a company or institution and holds the right position(s) will try to adjust the rules here and there in ways that they think are conducive to the goals of the reorganization. And they will do their best to install the right people in the right positions beforehand and will try to mold those minds for the reorganization and restructuring through all kinds of formal and informal influence. Anyone who experiences this up close at a company or institution will probably not experience this as a conspiracy. We could even say that every biological organism does the same: It tries to adjust its environment in the desired direction.

At certain points, however, the aforementioned practices may turn into something that does have the structure of a conspiracy. Large institutions do use all kinds of questionable strategies to impose their ideals on society, and the means to do so have increased spectacularly in recent centuries. The whole mechanization, industrialization, "technologization," and "mediatization" of the world has indeed led to the centralization of power, and no sane person can deny that this power is pursued in a relentless way, with a radical lack of ethical and moral awareness. It is well documented: Whether in governments, the tobacco industry, or the pharmaceutical lobby, there is bribery, manipulation, and fraud. Who doesn't partake in these practices can hardly remain at the top.

In their endeavors to impose their ideals on society, institutions and people do indeed cross ethical boundaries, and when this goes far enough, their strategies may indeed devolve into a full-fledged conspiracy: a secret, intentional, planned, and malicious project. It is also well known that, as the process of totalitarianization continues, the totalitarian regime is increasingly organized as a full-fledged "secret society."[18] For example, we have seen that the Holocaust came about through a mind-boggling process of mass formation that blinded both the perpetrators and the victims and drew them into an infernal dynamic (see chapter 7). However, at a certain level, there was also an intentional plan, which systematically aimed to optimize racial purity through sterilization and elimination of all impure elements. There were approximately five people who neatly and systematically prepared the entire Holocaust destruction apparatus, and they managed to make all the rest of the system cooperate with it in total blindness for a long time. And those who did see what was going on—namely that the concentration camps were in fact extermination camps—were accused of being . . . conspiracy theorists.[19]

The preparation and implementation of such plans are by no means the exclusive privilege of totalitarian regimes. Throughout the twentieth century, large numbers of men and women whose genetic material was considered "inferior" have been secretly sterilized under the doctrine of eugenics. By 1972, the term *eugenics* had taken on a too-negative connotation and was replaced by *social biology*, but the practice remained the same and continued into the twenty-first century (for example, the sterilization of California inmates without informed consent).[20] Do we have good reason to believe that, in recent years, such practices have ceased?

The fact that, in the current social climate, there is hardly any latitude to expose this decay in the exercise of power is highly dangerous. This is precisely the detrimental influence of the rise of the masses: It is so radically intolerant of dissent opinions that it labels any analysis of dangerous influence from institutions, companies, and so on as "conspiracy theory." *La passion de l'ignorance* (the passion for ignorance) is flourishing like never before. And paradoxically, fanatical conspiracy thinking contributes to this problem because it makes more nuanced analyses less visible and more prone to stigmatization. They are tarred by the same brush and guilty by association.

This makes it difficult for everyone to assess the presence and extent of malicious manipulation. Either it is completely ignored or it is perceived to be everywhere. The appeal of these two opposites can always be situated on an affective-impulsive level; both interfere with an authentic, sincere intellectual passion to want to know the truth. In the end, it is usually only a small group of people who manage to escape these forces and are able to make more nuanced and subtle assessments.

This gives rise to a polarization in society, which becomes divided between two camps: a large group (the crowd), who believes everything that appears in the mainstream media, however absurd it may be; and then another group, who completely distrusts the same story. Just as in Edgar John Rubin's famous drawing (see figure 8.2) in which one can see either a vase or two faces, but never both at the same time, these two groups perceive in the social developments a different picture of reality, a different gestalt, and cannot imagine that the other group perceives a totally different picture.

Figure 8.2.

The risk of a violent confrontation between these two groups is not nonexistent. Conspiracy thinking itself can also give rise to the emergence of a mass phenomenon. The famous witch hunts of the Middle Ages, leaving some cities and towns with hardly a woman alive, are examples of such phenomenon. And conspiracy theories such as *The Protocols of the Elders of Zion* also played an important role in the rise of the anti-Semitic masses of the Middle East and Nazi Germany. Nazi propaganda mimicked *The Protocols* in many ways; Heinrich Himmler and Adolf Hitler knew them by heart.[21] The narrow, causal attribution of all suffering to a small Jewish elite was adopted by the Nazis. This causal reasoning was a monstrosity in itself, but the absurdity inherent in the masses ensured that it was not so much the supposed Jewish elite, but millions of ordinary Jews who fell victim to it.

In this way, conspiracy thinking can be a reaction to mass formation, an interpretation of it, but it can also give rise to mass formation itself. It is not expected, however, that current conspiracy narratives will lead to large-scale mass formation. In 1951, Arendt already foreshadowed that the masses of the future would be dull, bureaucratic, and technocratic in nature.[22] Nowadays, certain conspiracy theories such as QAnon lead to small-scale mass formation, as we saw to a certain extent during the storming of the US Capitol. In this way, a small crowd can come face to face with a large one. However, in a physical confrontation, the smaller crowd will lose out. In doing so, it testifies in its own way to the blindness and, above all, the self-destructiveness inherent to mass formation. If one wants to slow down the masses, one must do so primarily by psychological means (discussed later in this chapter). Physical violence, on the other hand, will mainly incite the masses and make them more fanatically convinced of their righteousness and their sacred duty to persecute and destroy the minority.

For this reason, conspiracy thinking is something to be dealt with carefully, on an intellectual level as well as on an ethical and pragmatic level. It often arises as an explanation for the phenomenon of mass formation, but it shows a tendency to drift off course into theories that are increasingly distant from a nuanced view of reality and, on a psychological level, often lead to simplistic and caricatural views. Arendt gave a moderate and, in all respects, sensible answer to the question of what extent mass formation and totalitarianism can be traced back to a conspiracy: There is a certain conspiracy dimension in most social upheavals—those in power may even have little choice but to contrive things behind closed doors—but it is easily overestimated. If anything rules from the behind the scenes, it's not so much secret societies, but ideologies. There is a steering and organizing body, but it does not primarily consist of a conspiracy elite that manages the world in a planned and coordinated way, but rather of a typical way of thinking, an ideology. To put it in the words with which Charles Eisenstein rejected a one-sided interpretation in terms of conspiracies: "Events are indeed orchestrated in the direction of more and more control, only the orchestrating power is itself a zeitgeist, an ideology . . . a myth [and not a conspiracy]."[23] Such a consideration never attributes the cause of social dynamics to one single point. The whole of society has a part in its rise in one way or another; every person bears a responsibility in it. That's why this nuanced statement is usually unsatisfactory for those who thirst for certainty and seek to vent anger and frustration by pointing out one main culprit.

* * *

In the previous three chapters, we have discussed the psychology of mass formation and totalitarianism theoretically. At this point, it is useful to pose the question: Can we also do something with this theory in practice? Our analysis mainly highlighted the complexity of the phenomenon; to explain it in terms of a large-scale conspiracy doesn't help us any further. For this reason, we have to conclude that, first and foremost, the problem cannot be solved by the violent elimination of an evil elite. The essence of the problem of totalitarianism lies in enormous mass dynamics. This means the elimination of totalitarian leaders will be to no avail; they are utterly replaceable. This is how Arendt put it:

> *In substance, the totalitarian leader is nothing more nor less than the functionary of the masses he leads; he is not a power-hungry individual imposing a tyrannical and arbitrary will upon his subjects. Being a mere functionary, he can be replaced at any time, and he depends just as much on the masses he embodies as the masses depend upon him.*[24]

The leader is, so to speak, just the apex of the pyramid of the mass movement, and if he is eliminated, he will be replaced without the system destabilizing.

Violence as a reaction against mass formation and totalitarianism is, of course, effective when carried out by external enemies of a totalitarian system—for example, the war of the Allies against Nazi Germany—but it offers few prospects for internal resistance and is generally counterproductive. When the opposition uses violence, the crowd merely sees justification and a "get-out-of-jail-free" card to unleash its already enormous potential of frustration and aggression and take it out on those it views as the enemy (those who do not go along with the New Solidarity).

Arendt noted that nonviolent resistance, on the other hand, is remarkably successful against totalitarianism.[25] She comes to that conclusion on the basis of historical observations—for example, the effectiveness of the resolute refusal of the Danish government and population to participate in the anti-Semitic measures that the Nazis tried to impose, but she fails to offer a psychological explanation. We can do that to some extent on the basis of the psychological description we have provided thus far. Furthermore, we can also describe the idea of "nonviolent resistance" in a more refined way.

Both the masses and their leaders are gripped by an ideologically colored narrative, the masses are hypnotized, the leaders are under a form of self-hypnosis. Both, so to speak, are in the grip of a *voice* (see the importance of indoctrination and mass media propaganda described in chapter 6). Mass formation, as a form of hypnosis, is a phenomenon where individuals are in the grip of the resonance of a voice—the voice of the leader of the crowd. However, not all of the population falls prey to this process. In chapter 6, we identified three groups that form when a mass rises: the masses themselves, who truly go along with the story and are "hypnotized" (usually about 30 percent); a group that is not hypnotized but chooses to not go against the grain (usually about 40 to 60 percent); a group that is not hypnotized and actively resists the masses (ranging from 10 to 30 percent).

The first and foremost guideline for members of this third group is that they should let their voices be heard and in as sincere a way as possible so as to not let the resonance of the dominant, hypnotic voice become absolute. The way in which this can happen varies throughout the process of totalitarianism (the dissident voice is progressively more censored and banned from mass media and from the public sphere), but there always remain opportunities. The assertion of a different voice always has an effect on the other two groups. As Gustave Le Bon described in the nineteenth century, dissonant voices (i.e., the voices of the third group) usually do not succeed in breaking through the hypnosis of the first group, but it does reduce the depth of the hypnosis and prevent the masses from committing atrocities. Also, the leaders prove sensitive to the dissonant voices, which is what we described in the previous chapter where we referred to the "waking up" of the Nazi leaders who were deployed to Denmark and Bulgaria. Asserting one's voice should typically be done in the calmest and most respectful way possible, never in an intrusive way and always with sensitivity to the irritation and anger it may generate but with determination and persistence. Although the dissident voice typically provokes rejection, and under certain circumstances also aggression, it is worth realizing that the masses also need this in order to not fall prey to themselves. We described this in chapter 7: If the opposition is silent, the totalitarian system becomes a monster that devours its own children. For this reason, it is an illusion to think that silence is the safest option, from whomever concerned.

The dissident voice also has an effect on the second group, the group that is compliant but not hypnotized. In contrast to the first group, this group is responsive to the quality of rational argument. Therefore, it is important that the dissident voice analyzes and refutes the indoctrination and propaganda of the totalitarian narrative in the clearest and most substantiated way possible. In a sense, this isn't difficult since the totalitarian discourse, especially its typical excessive use of numbers and statistics, is usually simply absurd. For the opposition, it is a matter of repeatedly and persistently, through the (limited) channels available for that purpose, piercing the web of appearances and showing, insofar as possible, the way in which a false image is being created. It is important to note that the counterargument should never aim at reversing the process of mass formation and a return to the prior prevailing state ("the old normal") because this is precisely the environment from which mass formation arose—from a profound psychological unease and suffering, which I described in chapter 6 (the four psychological conditions for mass formation). Attempting to convince people to return to this is completely nonsensical and will provoke the opposite effect: Those who are in the grip of the mass formation will cling even more stubbornly to their narrative. In general, counterarguments should be formulated in a disciplined and organized manner, through a specially created structure of working groups, specialized in certain themes and topics. The formation of such groups, in itself, also provides an antidote to one of the most pernicious effects of totalitarianism: the destruction of every social bond and structure.

Finally, the third group speaks for itself. This group usually becomes, to a greater or lesser extent, the object of the frustration and aggression of the masses (see chapter 6). It is typically dehumanized, presented as creatures of inferior humanity. If this group ceases to assert its voice, it confirms the stigma. Speaking and rational reasoning is what distinguishes humans from animals; to stop speaking out paves the way for dehumanization. This in itself shows the importance of continuing to speak out as calmly and wisely as possible. But there is another important reason to do so. Speaking leads to experiences of meaning and existence, at least if the one who speaks tries to express his subjective truth as honestly and sincerely as possible. Dissident speech doesn't have to be primarily tactical or rhetorical in nature, but it should be authentic and honest (see chapter 7). Even if speaking out has no effect on the Other, it will still do something for oneself. Eventually, it is in this act of truth-telling that the absurdity of totalitarianism becomes meaningful: Those who do not join in the collective madness and quietly and sincerely continue to assert their opposing voice are, by doing so, steadily elevated in their humanity. Read, for instance, Solzhenitsyn's poignant testimony on the effects on himself that speaking out and writing had during his eight-year stay in the gulags.[26]

The first and foremost task is to keep speaking out. Everything stands or falls with the act of speaking out. It is in the interest of all parties. The specific manner in which the act of speaking out takes place—in books, publications or interviews, in front of the cameras, in shops or at the kitchen table, in the company of a limited or large group of people—is of less importance; everyone who, in his own way, speaks out about the truth, contributes to the cure of the ailment that is totalitarianism. It is not necessary to have a huge number of people who unite in speaking out to form a meaningful social group. Remember that the masses (the totalitarized portion of the population) usually consist of only about 30 percent of the total population, and the 40 or 50 percent who meekly follow do so mainly because the masses form the largest contiguous block and have the loudest voice, which to them is the most convincing. However, the absurdity of the discourse of the masses also plays to their detriment. If this remaining 10 to 20 percent can form a countergroup (without becoming a crowd themselves!) and is able to assert an alternative voice in a sensible way, this group will then be able to undo the mass formation, or at the very least, to free society from its grip. Moreover, the nonconformist group has to always bear in mind that the masses (and the totalitarian system) are intrinsically self-destructive and always destroy themselves in the long run (see chapter 7). The totalitarian system doesn't have to be overcome so much as one must somehow survive until it destroys itself.

A more strategic option to break through the mass formation could also be considered: replacing one object of anxiety with another. Mass formation occurs when free-floating, unbound anxiety attaches itself to an object of anxiety (see chapter 6). This connection can be undone if another object that instills even more anxiety is presented. For example, one could try to circulate an alternative narrative that puts the totalitarian regime itself forward as an object of anxiety (thus, evoking the atrocious consequences of totalitarianism). If at the same time, this story also offers a strategy to deal with that new object of anxiety, one could indeed achieve a more durable reorientation of the anxiety in individuals. This could work to some extent. If such a strategy is applied moderately, this amounts to a warning for a real danger with good reason. However, if this is made to be the primary strategy, whose entire focus lies on the instillment of anxiety, one crosses ethical boundaries and will drift into a dehumanization process, which is in no way different from the one that is typical for mass formation.

* * *

I have provided a few guidelines for defense against the psychological mechanism of mass formation. Of course, these guidelines in themselves are only superficial. The rise of the masses and totalitarianism is ultimately grounded in mechanistic thinking (as we discussed in the first five chapters of this book). For this reason, ultimately, we have to get beyond the mechanistic ideology in order to come to a substantive sociocultural solution. In the final three chapters, we will examine whether the mechanistic ideology has some openings that could offer us another vision of the world, and of mankind.

PART III

BEYOND THE MECHANISTIC WORLDVIEW

CHAPTER 9

The Dead versus the Living Universe

The following is broadly the causal reasoning we have presented in this book: The mechanistic ideology has put more and more individuals into a state of social isolation, unsettled by a lack of meaning, free-floating anxiety and uneasiness, as well as latent frustration and aggression. These conditions led to large-scale and long-lasting mass formation, and this mass formation in turn led to the emergence of totalitarian state systems.

Therefore, mass formation and totalitarianism are in fact *symptoms* of the mechanistic ideology. Just like an individual physical or psychological symptom, these social symptoms signal an underlying problem: In this case, that a large proportion of the population feels socially isolated and suffers from intense experiences of anxiety and meaninglessness. And just like individual symptoms, they generate a *disease gain*. For example, they transform the experiences of social isolation and fear into an illusion of connectedness. And as with individual symptoms, they generate this disease gain while failing to solve the underlying problem itself.

For this reason, we need an analysis of the underlying problem—that is, the cause of the symptom, namely the mechanistic ideology. Societies are primarily besieged by *ideas*. The most fundamental change that we as a society have to aim for is not a change in practical terms but a change in consciousness. In the first part of this book, we examined the psychological problems caused by the mechanistic ideology; in the final part, we will examine how we can transcend this ideology. In this chapter, we will reflect upon one of the core characteristics of the mechanistic ideology. This ideology sees the universe as a logically knowable, predictable, controllable, and undirected mechanical process. And above all, it sees the universe as a dead and meaningless given, as the blind, mechanistic interaction between dead, elementary particles. While such a view of the world and matter imposes itself as the only scientifically valid view, a thorough examination teaches us that, from a scientific point of view, this world view is actually outdated.

* * *

The mechanistic worldview is, in fact, as old as man himself, or at least, it was already present in what we usually consider the early days of Western civilization. In the era of the ancient Greeks, about 400 BCE, atomists such as Leucippus and Democritus were already defending the idea that the universe, in its entirety, was essentially a collection of mechanically interacting material particles. Those particles were already called *atoms*, which means "indivisible" or, more literally, "unsliceable" (*atomos*).

It was not until the Enlightenment, however, that mechanistic thinking became dominant and provided the only remaining Grand Narrative of Western culture. As we discussed in chapter 1, this ideology even furnished a kind of creation myth: Everything starts with a big bang that sets the machine of the universe in motion and, through a series of mechanistic effects, produces first a series of inorganic elements and subsequently also living beings. Within this reasoning, the world is a dead mechanistic process, an enormous chain reaction of collisions of elementary particles that continues endlessly, without purpose or direction, and somewhere along the way, randomly produces life and mankind.

This entire process is seen as strictly predictable. The French mathematician Pierre-Simon Laplace expressed this in perhaps the most direct way:

> *We ought then to regard the present state of the universe as the effect of its anterior state and as the cause of the one which is to follow. Given for one instant an intelligence which could comprehend all the forces by which nature is animated and the respective situation of the beings who compose it [. . .] it would embrace in the same formula the movements of the greatest bodies of the universe and those of the lightest atom; for it, nothing could be uncertain and the future, as the past, would be present to its eyes.*[1]

Most philosophers have considered such a worldview to be naive. Bertrand Russell, for example, argued in his Russell's paradox that there can never be an entity, however much computing power it has, that can have complete knowledge.[2] Such an entity would also have to have a complete knowledge of itself, and also a complete knowledge of itself as an entity possessing complete knowledge of itself, and so on to infinity. In the twentieth century, Werner Heisenberg also proved this concretely: One cannot speak of elementary particles in terms of certainty. The more accurately their position in time is determined, the more uncertain becomes their location in space. "Not only is the universe stranger than we think; it is stranger than we can think." (See Heisenberg's uncertainty principle.)[3]

These elementary building blocks of the universe—atoms—appeared to be both more complex and more elusive than previously thought. The more the researcher's hand tried to close itself around them, the more they slipped through his fingers. Rather than the tiny, massive spheres envisioned by the ancient Greeks, twentieth-century physics showed them to be swirling, energetic systems, patterns of vibration rather than solid matter. Yes, in the final analysis, they even turned out not to be material phenomena at all but rather to belong to the order of consciousness. The great physicists of the twentieth century believed them to be mere thought-forms, mental phenomena that respond to the consciousness of researchers (as we shall discuss further in the chapter 10).

We could of course delve deeper into the findings of quantum mechanics to further relativize the idea of a mechanistic universe. But the phenomena of which quantum mechanics speaks are situated in a dimension that most people will never have access to. Who will ever get a direct look at the subatomic world? In this respect, there is another field of science that offers better, more concrete perspectives, namely the complex and dynamic systems theory and the chaos theory. These theories deal with phenomena that everyone, in principle, can sensorily perceive and that illustrate the limitations of the mechanistic vision in an equally convincing way.

* * *

When Benoit Mandelbrot—a brilliant mathematician, considered one of the founders of chaos theory—joined IBM, he was confronted with the problem of noise that interferes with computer signals transmitted over telephone lines.[4] This noise occurred due to a series of external factors, such as humidity, irregularities in the material of the lines, and small electromagnetic disturbances that hampered signal transmission in an accidental and incalculable way. We can only assume that these factors acted in a random way and independently from one another and therefore, normally, there cannot be any consistency in the noise on the telephone lines.

Mandelbrot was not a person who believed what everyone else believed, however. He was bold enough to assume that there might be a pattern in the noise after all. "Just because it doesn't make sense doesn't mean it can't exist," he said. And he was correct. In the noise, he discovered a well-known mathematical pattern, known as Cantor dust. Anyone can easily reproduce this pattern by repeatedly dividing a line into three segments and omitting the middle segment each time.

The big question, of course, is the following: How is it possible that a series of random factors, manifesting independently, can lead to a regular pattern? How could it be that damage caused to a cable by, say, a screwdriver and the magnetic disturbances of a thunderstorm become part of the same pattern? It is as if all these accidental, mechanical disturbances are drawn into a stable and strictly mathematically ordered field in order to be stripped of any coincidence. James Gleick put it this way: "Life sucks order from a sea of disorder."[5] The noise on a telephone line seems *to organize itself.* In living organisms, we have—erroneously—come to consider this quality of self-organization to be normal. Living beings breathe air, and eat and drink, and all these disparate elements bring about the ordered pattern of their bodies. However, when this phenomenon manifests itself in the inorganic world, we perceive it as a perplexing phenomenon and contrary to the prevailing worldview (which it is).

Another example is the regularity of water droplets, dripping from a faucet, as demonstrated by Robert Shaw.[6] This is an example from everyday life, observable by anyone. A relatively simple mathematical procedure suffices to show that there is mathematical regularity in the lapse of time between the drops dripping down, which, when represented visually, produces beautiful organic patterns. In this case as well, we encounter the curious paradox that the moment a drop of water drips down is, on the one hand, caused by a series of disconnected, external factors—the surface tension of the water, the temperature, vibrations in the surrounding air, the texture of the faucet's rim. But on the other hand, it seems to follow a strict pattern. The reason all these unrelated factors lead to a consistent pattern is difficult, even impossible, to explain within a mechanistic worldview. Obviously, this pattern can be disrupted by certain interferences—for example, by intentionally blocking the mouth of the faucet with your finger. However, after the cessation of this interference, where it is difficult to determine in which way it differs from the other external factors, the system returns to its spontaneous equilibrium and the pattern reinstates itself.

Gleick had the following to say about it: "Those studying chaotic dynamics discovered that the disorderly behavior of simple systems acted as a *creative* (italics added) process. It generated complexity: richly organized patterns, sometimes stable and sometimes unstable, sometimes finite and sometimes infinite, but always with the fascination of living things."[7] Please, take note of the qualifications *creative* and *living*. This aspect of creation and life in matter was overlooked by the classical scientific approach.

More or less in line with these examples, fractal theory (a subdomain of chaos theory) showed an unsuspected, mathematical determinacy of sets of natural forms, such as those of leaves, plants, trees, sea sponges, algae. The best-known examples are perhaps seashell patterns studied by Hans Meinhardt;[8] the Mandelbrot set; and the spiral shapes determined by the Fibonacci sequence. This last determination is so simple that it is easily understandable, even to nonmathematicians. The Fibonacci sequence consists of a series of numbers that is obtained by starting with the numbers 0 and 1 and then continuing with a number that is the sum of the two previous numbers (so 0, 1, 1, 2, 3, 5, 8, etc.). This series of numbers determines the curves of a spiral that can be found everywhere in nature. Galileo's famous statement in 1623 that "The book of nature is written in the language of mathematics" must be taken literally, it seems.[9]

Let's take a closer look at one example. Lorenz's chaotic waterwheel is a mechanical device that makes movements that show direct similarities with the dynamics of convection patterns in liquid and gas. (See figure 9.1.) It was designed by MIT professor Willem Malkus in 1972 to illustrate the work of Edward Lorenz, a mathematician and meteorologist and one of the founders of chaos theory. It consists of a rotating wheel to which small buckets with a bottom hole are attached. At the top, there is a tap that provides water flow into the top bucket. At a very low influx, the wheel does not move, simply because the water flows out of the hole in the bottom of the bucket faster than it flows in. At a slightly higher influx, the bucket will fill up and the wheel will start to move, sometimes in one direction, sometimes in the other. Once the wheel has chosen a certain direction, the behavior of the wheel is regular and predictable and directly correlated with the influx of water: The greater the influx, the faster it turns.

Figure 9.1. Lorenz's water wheel

If the influx exceeds a certain limit, however, a series of complex effects occur that cause the wheel to behave erratically. The top bucket initially fills to the brim, causing the wheel to turn at a high speed. But then, because of the high speed, the other buckets hardly get a chance to fill up as they pass by the top. This causes the wheel to slow down and possibly come to a temporary stop, whereupon it continues to rotate in the same direction, or sometimes in the opposite direction. This process is repeated with countless variations; the wheel sometimes moves quickly, sometimes slowly, sometimes in the same direction for a prolonged period of time, sometimes constantly changing direction. The irregularity in the chaotic phase was shown to be total in nature. This means that there is no (strictly) repeating pattern or repeating *period* in the wheel's movements.

No matter how chaotic the movements appear, they surprisingly turned out to be strictly determined. They can be described by a mathematical model consisting of three iterative differential equations with three unknowns (which in themselves are actually a simplification of the much more complex Navier-Stokes convection equations). In conformity with the chaotic behavior of the wheel, the (endless) series of solutions of these equations shows no periodicity either. Or, in other words, there is no recurring pattern in the set of values of the unknowns generated by the equations.

Therefore, the dynamics of the wheel closely resemble the structure of irrational numbers, such as pi, whose digits after the decimal point do not show any periodicity either. The qualification of such numbers as "irrational" primarily refers to the fact that such numbers cannot be written as a fraction, as a *ratio*. However, in laymen's terms, "irrational" in the sense of *not rational* is not incorrect either. It is true that such numbers cannot be rationally envisaged. That makes them disruptive in a logically ordered, rational worldview. Hippasus (a follower of Pythagoras)—who is considered the person who discovered these irrational numbers—experienced this to his own detriment. Legend has it, he was on a ship with his brethren Pythagoreans and was promptly thrown overboard when he articulated his intuition that there exists something such as irrational numbers. This illustrates clearly: The limits of the ratio always lead initially to *uncertainty, fear, and aggression.*

The combination of chaotic behavior and determinism gives the waterwheel the fascinating property of "deterministic unpredictability."[10] It amounts to the following: Even with the waterwheel formulas at hand, it is not possible to predict, even only one second in advance, how it will behave. The reason for this is simple: To be able to predict how the waterwheel will behave in the future, you need to measure the wheel's state of motion in the present and enter it into the formulas. But due to the nature of the wheel, even immeasurably small differences in the current state of motion can lead to radical differences in future behavior (in systems theory, this is called the property of "sensitivity to initial conditions"). Therefore, the wheel continues to shroud its future in mystery forever.

What is most fascinating about the story of Lorenz's waterwheel is this: At some point, Lorenz got the idea to plot the successive values of the three quantities in the equations on a three-dimensional orthogonal coordinate system, also called *phase space* in chaos theory. Curiously enough, it was not just a random nebula of points that appeared, as one would initially expect with a chaotically behaving system. What emerged was a very regular figure with striking aesthetic features, which has since been known as the Lorenz attractor (see figure 9.2).

As Gleick said, "Phase-space portraits of physical systems exposed patterns of motion that were invisible otherwise, as an infrared landscape photograph can reveal patterns and details that exist just beyond

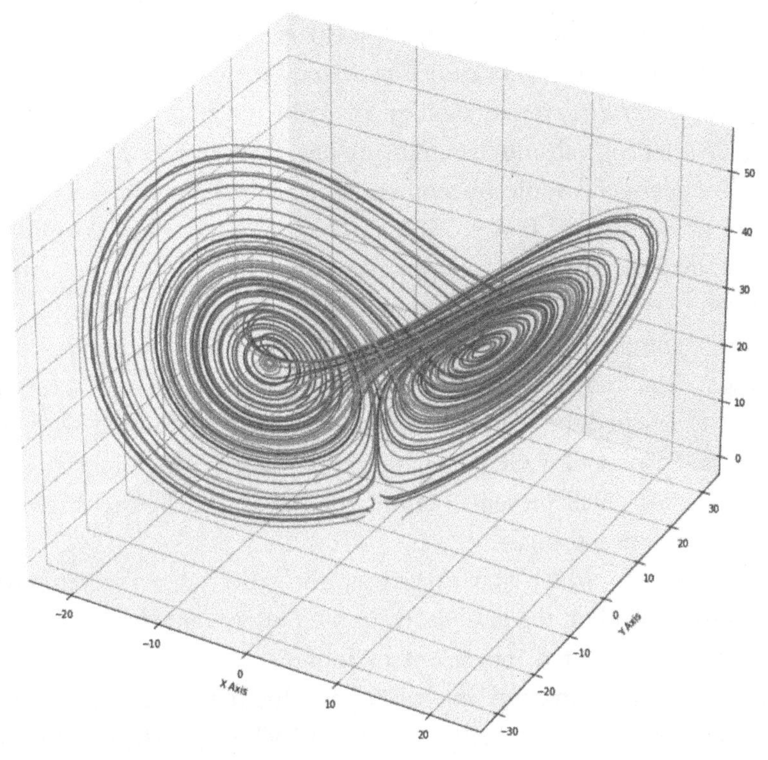

Figure 9.2. The Lorenz attractor

the reach of perception."[11] Lorenz was the first to show that certain chaotically manifesting behaviors are nevertheless determined by a strict (and sublime) order and can be visually represented in phase space. Hidden beneath the apparent chaos of the superficial experience of the wheel is an aesthetically magnificent order of universal forms, in many ways reminiscent of Plato's ideal world. The quantum physicists also arrived at Plato's famous ideal world, albeit via a different route. Heisenberg expressed this in perhaps the most direct way: "I think that modern physics has definitely decided in favor of Plato. The smallest units of matter are not objects in the ordinary sense; they are forms, ideas. . . ."[12]

This is without doubt the most important lesson that the waterwheel has to teach us: We cannot predict the specific behaviors of the waterwheel (at least not in its chaotic phase), but we can learn the principles by which it behaves and learn to sense the sublime aesthetic figures hidden beneath the chaotic surface of those behaviors. Hence, there is no rational predictability, but there is a certain degree of *intuitive* predictability. In 1914 already, Henri Poincaré argued that logical understanding is not always necessary to intuitively understand some phenomena and to make predictions based on one's intuition.[13] It is possible to accurately sense the globality of the underlying structure of a phenomenon—for example the Lorenz attractor—without having any significant logical understanding of that phenomenon. Poincaré even went a step further, stating that pursuing logical knowledge about the phenomenon might, once a certain point is reached, be counterproductive. When confronted with the irrational aspect of a phenomenon, the persistence to obtain rational understanding will prevent us from coming to conclusions based on more direct receptiveness.

The way in which you experience the wheel as a spectator will strongly depend on the level at which your attention is focused. If you look at each isolated movement or motion sequence separately, the movements are perceived as chaotic and disparate. The wheel seems like a cacophony of abruptly interrupted back and forth movements. However, if you are able to feel affinity with the wheel and get to sense the deeper rhythms present in the variety of movements (as represented in the figure of the Lorenz attractor), then you experience the timeless, creative harmony that is present underneath the variety of superficial movements and the wheel becomes an appeasing phenomenon.

In this respect, the wheel teaches us something that applies to a far broader extent to the human being, society, life, and nature. Just like the wheel, most phenomena in nature are complex and dynamic and, in their complexity, are rather unpredictable. But like the wheel, life follows certain principles and sublime phenomena are hidden beneath its seemingly chaotic surface. And this is perhaps a person's greatest task: to discover the timeless principles of life, in and through all the complexity of existence. The better we can sense those principles, the more we feel that we start to understand some of the essence of life and that we are connected with the majestic, ordering principle that speaks to us from across the universe. And the more we stick to our principles, even if it seems to our own detriment in the short term, the more real these principles become and the more we develop, as human beings, a real sense of existence and fortitude. Being too opportunistic and relinquishing our principles because "smart" analysis of a situation suggests it might be advantageous, often leads to a loss of individuality and experiences of meaninglessness. If one focuses too much on the superficial appearances of life and loses touch with the underlying principles and figures, life will increasingly be experienced as a meaningless chaos, just like Lorenz's waterwheel.

The same applies at the societal level: A society primarily has to stay connected with a number of principles and fundamental rights, such as the right to freedom of speech, the right to self-determination, and the right to freedom of religion or belief. If a society fails to respect these fundamental rights of the individual, if it allows fear to escalate to such an extent that every form of individuality, intimacy, privacy, and personal initiative is regarded as an intolerable threat to "the collective well-being," it will decay into chaos and absurdity. The belief in the mechanistic nature of the universe and the associated overestimation of the powers of human intellect, typical of the Enlightenment, were accompanied by a tendency to lead society in a less and less principled manner. Within a purely mechanistic way of thinking, it is extremely difficult (not to say impossible) to ground ethical principles. Why should a machine man in a machine universe have to adhere to principles and ethical rules in relationships with others? Isn't it ultimately about being the *fittest* in the struggle for survival? And therefore, aren't ethics and principles a hindrance rather than a merit? In the final analysis, it was no longer a question for Enlightenment people to adhere to commandments and prohibitions or ethical and moral principles, but to move through this struggle for survival in the most efficient way possible based on "objective knowledge" of the world. This culminated in totalitarian and technocratic forms of government, where decisions are not made on the basis of generally applicable laws and principles but on the basis of the analysis of "experts." For this reason, totalitarianism always chooses to abolish laws, or fails to implement them, and prefers to rule "by decree." This means that, each new situation will require the formulation of new rules on the basis of a (pseudo)rational assessment of such situation. History abundantly illustrates that this leads to erratic, absurd, and ever-changing rules, which ultimately destroy all humanity in society.

This is perhaps the most direct and concrete illustration of Hannah Arendt's thesis that ultimately totalitarianism is the symptom of a naive belief in the omnipotence of human rationality. Therefore, the antidote to totalitarianism lies in an attitude to life that is not blinded by a rational understanding of superficial manifestations of life and that seeks to be connected with the principles and figures that are hidden beneath those manifestations.

Chaos theory and the complex and dynamic systems theory open a breathtaking new perspective on the universe. In his widely acclaimed book *Chaos*, Gleick states that chaos theory is the third great scientific revolution of the twentieth century (after the relativity theory and quantum mechanics).[14] Mechanistic-materialistic science started from the assumption that the world is logical and predictable and, in particular, that it essentially is a dead mechanical process. Science aimed to reduce living phenomena—the organic, the consciousness, etc.—to dead processes (for example, to mechanical chemical processes). Quantum mechanics and chaos theory shake this worldview. They initiated the reverse momentum and lean much more toward a vitalist worldview. They suggest that there is life and consciousness in all kinds of phenomena that we previously considered to be dead, mechanical processes. Think of the noise on telephone lines: It proved to not be the passive effect of all kinds of mechanical factors, but to be self-organizing; it is characterized by purposefulness and a sense of aesthetics.

Perhaps the most revolutionary aspect of chaos theory is that its observations allow us to see that there is indeed a final and formal cause at work in nature. These concepts are derived from the causality theory of Aristotle and are indispensable when considering the process of causation. In a nutshell, this theory states that there are four kinds of causes: the material, the efficient, the formal, and the final. Aristotle illustrated the difference between these four causes using the metaphor of making a statue. The material cause of the statue is the matter from which it is made (without such matter, no statue). The efficient cause is the movements of the sculptor, who uses chisel and hammer to transform the stone into a statue. The formal cause is the idea or form of the statue as it has taken form in the mind of the sculptor and determines how he will direct his movements. The final cause is the intention to make a statue (for example, because someone has ordered a statue from the sculptor). It is clear that, within a mechanistic worldview, only the material and the efficient cause are considered to be active. Once upon a time, the mechanistic universe, as a collection of material particles, set itself in motion, and all the rest followed from the initial movement of the particles. So the particles in themselves are the material cause; their movements, which generate all kinds of effects, are the efficient cause. However, within such a worldview, it cannot be presumed that certain "forms" or "ideas" exist in advance (those of certain organisms, for example) that would influence the way the material process unfolds.

Chaos theory proves that such forms *do* exist and that they operate in a coordinated manner. What has been demonstrated with the noise on telephone lines and drops dripping out of faucets can be broadened to a much larger scope. Chaos theory shows us that the mountain landscape that transports us in breathless admiration is not simply the effect of a lifeless mechanistic process—accidental mechanistic processes between tectonic plates, erosion, and eruptions of lava—but that a timeless and sublime idea coordinated the myriad of mechanical processes involved in its formation. Chaos theory heralds, maybe even more than quantum mechanics, the era that historically and logically follows the Enlightenment; an era when the universe is once again pregnant with meaning.

CHAPTER 10

Matter and Spirit

The first basic assumption of the mechanistic-materialistic worldview is that the universe is a machine-mechanistic given that can be fully understood by means of logical reasoning. In the previous chapter, we discussed the relativity of this theorem. In this chapter, we will tackle the second great assumption of mechanistic materialism: Everything belonging to the domain of consciousness and the psychological realm is a consequence of material phenomena—*matter over mind.*

Contemporary public discourse shows a certain ambiguity when it comes to the psychological dimension of being human. On the one hand, psychological well-being is considered to be of crucial importance. It is believed that stress has adverse health effects, it is recognized that placebo effects play a major role in medical interventions, it has become more or less commonly recognized that it is important to "talk about our problems," and so on.

On the other hand, the world is still firmly in the grip of the mechanistic view of the world and mankind. Maybe even more than ever before. Within this ideology, everything belonging to the domain of consciousness and the psychological experience is ultimately considered an insignificant by-product of the biochemistry of the brain. Man's desires and aspirations, his romantic longings and his most superficial needs, his joys and his sorrows, his doubts and his choices, his pleasures and his sufferings, his deepest aversion and his most lofty aesthetic appreciations—in short, his complete subjective world of experience—is reduced to a consequence of elementary particles in his brain that interact according to the laws of mechanics.

Obviously, such a viewpoint has to consider any psychological approach to life—and by extension any religious or spiritual practice—as a form of irrationality. And any therapeutic application of such conceptual frameworks is, at best, designated as a *temporary Band-Aid*, a fringe therapy that may be tolerated until a real, *biological* treatment is discovered that will address the real, biological cause of human suffering. Depression originates in the brain, and if we look hard enough, one day, we will be able to show clearly which mechanical error is its underlying cause and, at that point, mechanistically repair such glitches in the machine.

Within such a worldview, one implicitly or explicitly assumes that there is a hierarchy in the sciences. The most fundamental level is that of physics, of the mechanistic interactions between the elementary particles, and everything else merely follows from this process. Physics determines inorganic chemistry; inorganic chemistry determines organic chemistry; organic chemistry determines anatomy and physiology; anatomy and physiology determine psychology; psychology determines economics, politics and sociology (see figure 10.1). Ultimately, everything can be traced back to physics and chemistry.

As widespread as this worldview is and as compelling as it may be in its simplicity, science has actually rendered it obsolete. First of all, quantum mechanics, as a science of elementary material particles, showed that it makes no sense to try to fully explain the domain of consciousness at the level of material knowledge. To a certain extent, elementary particles *themselves* are determined by the domain of consciousness—for example, by the mental act of perception during experiments. As inconceivable as this may seem, it is a fact that, if a particle is observed by two people at the same time, this same particle can be in two places at the same time.

Moreover, not only is the momentary localization of the particle determined by the observation but also the entire trajectory traveled by this particle in the billions of years prior to the moment of the observation.[1] It is only at the time of observation that the past trajectory is determined. According to world-famous physicist Stephen Hawking, "The choice [of a particle] whether to take one or both paths in this case would have been made billions of years ago, before the Earth or perhaps even our sun was formed, and yet with our observation in the laboratory we will be affecting that choice."[2] These insights are so contrary to the way in which we experience and understand time, space, and matter that the human mind is hardly able to grasp them. Niels Bohr expressed the strangeness of the observations of quantum mechanics as follows: "Anyone who is not shocked by quantum theory has not understood it."[3]

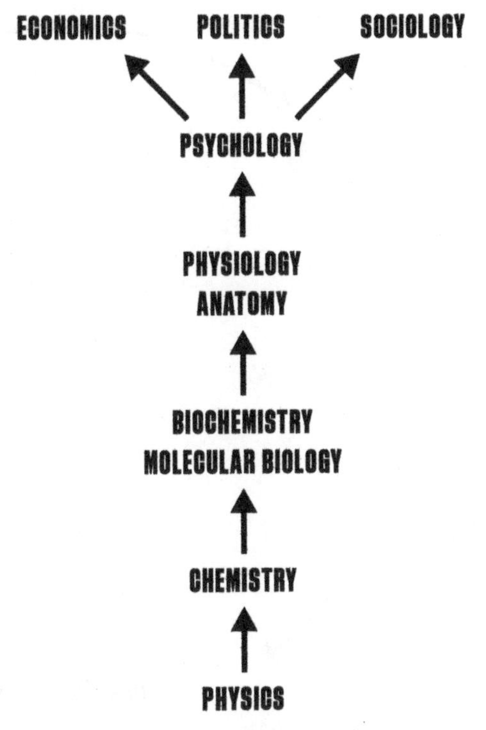

Figure 10.1. Hierarchical organization of the sciences within a strictly mechanistic-materialistic way of thinking

For this reason, we see that this hierarchy in the sciences, where the material domain determines the realm of physics, the realm of psychology is not universally valid: Man as a psychological being equally determines the domain of material objects. Therefore, we have to at least assume a mutual influence or circular causality between consciousness and matter (figure 10.2). The founders of quantum mechanics went even much further and considered the material domain to essentially form part of the realm of consciousness. As Werner Heisenberg says: "In fact the smallest units of matter are not physical objects in the ordinary sense; they are forms, ideas."[4] The logical positivist philosopher

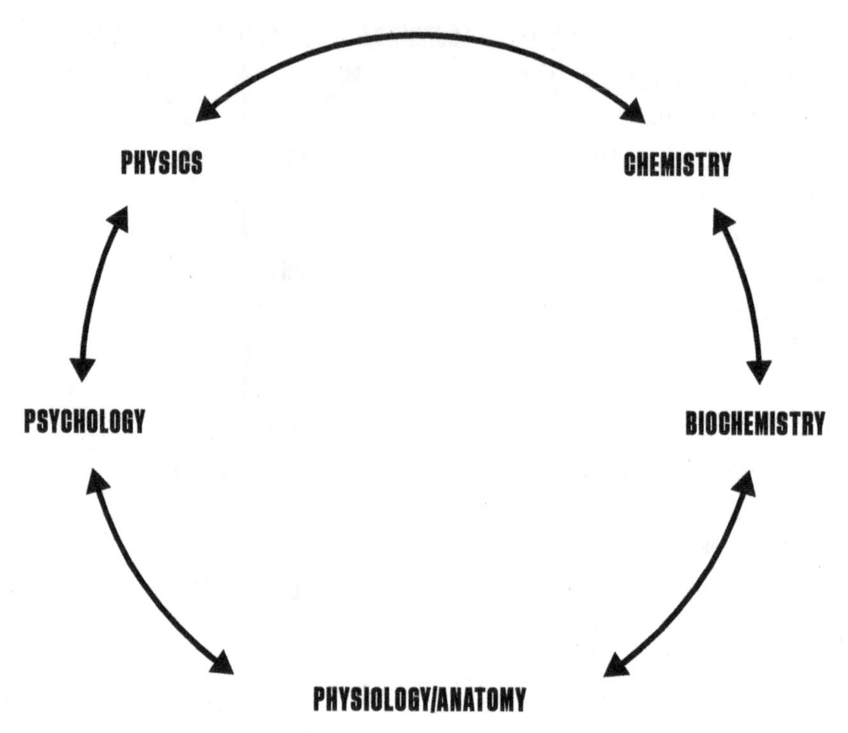

Figure 10.2. Circular causality between the different scientific domains

Bertrand Russell also took the same view: "All our data, both in physics and psychology, are subject to psychological causal laws. . . . In this respect psychology is nearer to what actually exists."[5]

The mechanistic worldview is based entirely on the idea of material particles as solid, absolute, "objective" data from which everything else can be inferred. But quantum mechanics shows us something radically different. The more intricately one examines matter, the more the act of observation itself influences the perception, and thus, the more subjective the perception becomes. Fully in line with Heisenberg's uncertainty principle, we can therefore state that matter—once regarded as the rock-solid foundation of mechanistic materialism—turns out to be essentially a subjective phenomenon. What exactly is matter? Nobody knows.

For this reason, a complete understanding of the materiality of the brain will never lead to a complete understanding of consciousness.

Every study of the brain as the material basis of consciousness will, at some point, encounter an absolute limit beyond which consciousness itself starts to determine matter. This shows that the psychological realm is a primary dimension, which, under no circumstances, can be reduced to the physical, the chemical, or the biochemical domain. At the therapeutic level, this also means that psychological treatments can indeed be full-fledged causal treatments.

* * *

Quantum mechanics may be the most fundamental refutation of the illusion of mechanistic-material determinacy of psychological experiences, but it is not the most concrete. There are observations that show in a more direct way that the psyche is difficult or impossible to reduce to a mechanistic brain apparatus.

For example, there are people in whom almost all brain tissue has died, who sometimes have less than 5 percent left, but whose mental functioning is still completely normal and who, for instance, score higher than 130 on an intelligence test. For the sake of clarity, I am not talking about obscure assertions but about scientific observations reported in journals such as *The Lancet* and *Science*.[6] Imaging or autopsy showed unequivocally that the brain cavity in such people was almost completely filled with fluid (see figure 10.3).

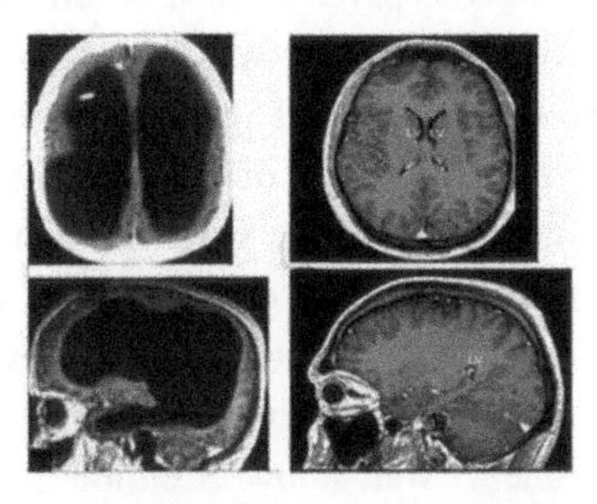

Figure 10.3. Comparison between scans of a person with intact brain tissue (*right*), and a person whose brain tissue has almost completely died off (*left*) but who still functions fine mentally. The black areas in the left skull show how the space released by the died-off brain tissue has been filled with fluid.[7]

In principle, these observations do not exclude a biological determinacy of consciousness. They only show that such a determinacy, if it does exist, has to be extremely complex in nature, and that the brain—to speak in terms of complex and dynamic systems—at least possesses the property of self-organization and self-reorganization. The little remaining brain tissue appears to have spontaneously taken over the functions of the dead brain tissue. However, such reorganization in and of itself presupposes a certain form of consciousness and intention in the brain tissue. Therefore, the hypothesis that consciousness is strictly determined by the material substrate of the brain ends up in circular reasoning: Consciousness is an effect of the material functioning of the brain, the material functioning of the brain is (to a certain extent) an effect of consciousness.

Along those lines, there are also the experiments on so-called neuroplasticity. Mental exercise (e.g., mathematical or memory training) leads to observable changes in the biochemistry and architecture of the brain, even in the relatively short term.[8] This also shows that the causal relationship between consciousness and brain is not a one-way relationship.

* * *

We could also refer to a series of observations that show in a direct way that the psychological realm can be the cause of the physical realm, rather than vice versa. Some of those observations can be made from occurrences that present themselves abundantly in everyday life. The most mundane, of course, is how certain emotions affect our body, or how human hair can turn completely gray in a few hours under the influence of, for example, intense fear or sadness. Or in a positive sense, we could refer to events where people gain an almost unimaginable strength, in circumstances where it empowers them to save a loved one. A well-known example is the story of Laura Schultz, a 63-year-old grandmother from Florida, who, in 1977, was able to lift the front wheel of a school bus with one hand to pull her grandchild out from underneath the bus with her other hand.

Such examples should open our eyes and convince us that we need to devote a lot more effort to better understanding psychological experiences. Strangely enough, in many cases man is blind to such direct evidence from his own experience and is more easily convinced by what "scientists" have "observed," although, in the latter case, he has to depend on "blind belief." Be that as it may, I'm happy to present some scientific findings in this matter.

* * *

The field of so-called psycho-neuroimmunology allows us to estimate the role that anxiety and stress play in the course of viral infections (which, of course, is not without relevance to the coronavirus crisis). Several studies report that mice are 40 percent more likely to die from viral infections due to experimentally induced stress.[9] The working mechanism is known: The stress leads to a reduction in immunity (mainly due to changes in hormone and white blood cell concentrations) and thus to greater susceptibility to viruses. In 2016, a study confirmed that the same mechanisms are also at work in humans and have a significant impact on mortality rates in a variety of severe physical conditions.[10] Important for the coronavirus crisis is a report from 2008 that stress leads to a higher mortality rate, especially in viral lung diseases, and that this effect is significantly greater in men than in women.[11] This corresponds with the difficult-to-explain observation that there are more male than female fatalities in the coronavirus crisis.

Observations are also being made in other fields of science that require neither statistics nor animal testing to convince us of the deadly nature of anxiety. It is well known in anthropology that in so-called primitive societies, people sometimes die after a shaman casts a curse on them. Herbert Basedow describes the typical course of such a ritual, as performed among Aboriginal peoples in Australia:

> *A man who realizes that he is being pointed out with the magic bone makes a pitiful impression. He is perplexed, his eyes staring at the dreaded bone, he stretches out his hands as if to stave off a deadly force that seeks to penetrate his body. His face turns white and his eyes become glassy, the expression on his face is hideously distorted, like a person suddenly paralyzed. He tries to scream, but usually the sounds get stuck in his throat and all that happens is foam forming on*

> *his lips. His body begins to tremble and his muscles contract uncontrollably. He flips backwards, falls to the ground and appears to swoon for a moment; but he soon begins to writhe as if in agony, covering his face with his hands, he begins to moan. After a while, he recovers somewhat and crawls to his cabin. From then on, he withers away and becomes increasingly ill, refusing to eat and taking no part in any of the tribe's daily activities. Unless there is another shaman to undo the curse, he will die within a short period of time.*[12]

This type of demise has been widely observed and, in the literature, it is known as a *psychogenic death*. Henry Ellenberger additionally specifies that it is important that *the entire community* to which the shaman and victim belong believe in the authority of the shaman. We'll come back to that later.

This may apply to an irrational primitive person who failed to outgrow magical thinking, but surely not to a rational Western person in the twenty-first century? Nothing could be further from the truth. There are countless observations that show that Western man, in his physical functioning, is equally subject to such phenomena. Professor Marie-Elisabeth Faymonville, anesthetist at the University Hospital of Liège, has been performing surgeries on patients under hypnosis for decades. The procedure, which was shown in a documentary on Belgian national television, looks astonishingly uncomplicated. Faymonville speaks to the patient, who is lying on the operating table, in a soothing way, leads him along in a relaxing mental world, and then gives an unobtrusive sign to the surgeon, conveying that he can start with the surgery. The surgeon is then easily able to make the required incisions in the body and carry out the prescribed medical procedures without the patient being aware of anything. And let's be clear: It does not only concern minor interventions, it also involves procedures like the surgical removal of the thyroid gland, the placement of breast prostheses, or the removal of tumors.[13]

As a matter of fact, such phenomena take place on a daily basis and en masse in medical practice, in the form of what became known as the *placebo effect*. This term came into vogue after curious observations were made on the battlefields of World War II. When doctors ran out of morphine, one of them came up with the idea to reassure soldiers just before an amputation by giving them an injection with a saline solution. To their surprise, they noticed that most soldiers were equally sedated, as if they were given morphine. Since then, there has been a growing body of research showing that placebos are capable of the most amazing physical effects, from opening coronary arteries in angina pectoris to reactivating dead areas of the brain. Authors such as Arthur Shapiro and Bruce Wampold, experts in the field, believe that the placebo effect accounts for the lion's share—often more than 80 percent—of the effects of medical interventions.[14] Some researchers, whether sensibly or not, argue for an almost generalized use of placebos instead of real medication. I should mention that other researchers arrive at substantially lower estimates on the basis of statistical research (some report only a 10 percent placebo effect).[15] Maybe this should all lead us to the conclusion that numerical research has to be put into perspective. The simple case studies of sedation via hypnosis and saltwater anesthesia are ultimately more scientifically valuable and leave no doubt that the impact of psychological factors on the body is, at least under certain circumstances, no less than phenomenal.

The placebo effect shows us the enormous importance of the patient's subjective experience of a therapeutic intervention. If someone is positive about the intervention, that in itself is an important part of the cure. But the reverse also applies: If someone has a negative attitude toward a treatment, this may have negative effects. This is known as the *nocebo effect*. There is extensive literature suggesting that a diverse range of conditions may be generated by this effect.[16] The psychogenic death, discussed above, is an extreme example and it shows that these effects can also be extremely strong. This shows us that, in addition to ethical reasons, there is a pragmatic, intellectual argument to never make medical treatments mandatory and to strictly enforce the right to self-determination.

On closer inspection, the mechanism of psychogenic death, hypnotic sedation, and placebos is the same every time: An authority figure evokes a powerful mental image in the individual who is being addressed. This mental image can be positive (e.g., healing) or negative (e.g., dying, becoming ill), but it has to be vividly and clearly present in the experience and it has to draw the attention away from all other mental activity. Thereupon, the body, so to speak, "merges" with that mental image, and the body takes the form or condition of this mental image (i.e., it gets well, it dies, it gets sick).

The curious, far-reaching influence of mental images on the body may have most convincingly been demonstrated by biologists. Harrison-Matthews showed in repeatedly replicated experiments that the ovary of female pigeons does not mature if the pigeon never sees the image of a congener (particularly if bred in complete isolation in a cage).[17] Follow-up experiments showed that it suffices to put up a mirror in the pigeon's cage for the pigeon to become fertile (albeit slightly less than when a pigeon grows up in the presence of a real congener). Rémy Chauvin did similar experiments with grasshoppers with even more far-reaching effects: As with the pigeons, there were strong influences on the functioning of the organs, but the color patterns on the shells were also different (there were no green stripes) and the anatomy of the rear legs differed consistently.[18] All kinds of variations on these experiments were carried out and each time the conclusion was the same: The decisive factor is whether or not visual images are present in the experiences of the animal in question.

What is important for the coronavirus crisis is the following: Several authors (e.g., Gustave Le Bon) have pointed out that the beliefs of a crowd (the group of individuals who identify with one another) have the same influence on the body as hypnosis. When society as a whole is in the grip of anxiety and the accompanying images of illness and death, those images in themselves become a causal factor. As described above, this happens in part because psychological distress radically changes the biological environment in which the virus enters by diminishing the immunity of such environment. Also think of the statement by Antoine Béchamp, which Louis Pasteur also endorsed at the end of his life: "The microbe is nothing, it is the environment that counts."

In this chapter, we mainly focused on the impact of visual images on the body. However, these images themselves inextricably form part of an even more important psychological register: the register of stories and ideologies, the *symbolic register*. The way in which narratives hold man and society in their grip is simply astonishing and also largely misunderstood. As we have already described in chapters 3 and 4, every child is involved in linguistic processes from an early age. It grows up within a narrative provided by the parents, which is usually shared by wider social groups and ultimately by an entire society. At its core, such a story always takes the form of a myth that provides a symbolic answer to the unanswerable questions. It provides a certain perspective on life, explains what is important and less important, determines what brings peace and what strikes fear. In fact, ethnographers like Marcel Mauss showed us that it determines even much more than that.[19] It determines what you will like and what you will reject (e.g., fish eyes are a delicacy in Congo, but are usually found to be repulsive in Europe), how the body moves (compare the gait of the Japanese people with that of African people), which basic reflexes we adopt when in pain (for example, the manner of pulling back a hand when in pain differs between cultures), and so on. It is no exaggeration to state that our bodies are completely absorbed and colonized by the mythical narrative in which we grew up.

This is why a medical procedure that works solely or mostly with words and narratives can have such enormous effects on the body. One can read texts such as *L'efficacité symbolique* by the great Belgian-French anthropologist Claude Lévi-Strauss to ascertain the enormous hold that symbolic structures have on societies, and on individuals who make up those societies.[20] It controls both mental and physical functioning in great detail. For example, Lévi-Strauss describes how, when a woman was experiencing complications during childbirth, shamans in the Brazilian rainforest repeatedly induced birth through a ritual that used an established tribal text that was read or sung in a ritualized way to the woman in labor. The text featured a series of characters from tribal mythology and told how a number of good spirits made their way through a narrow corridor leading to a cave where evil spirits imprisoned the baby. The good spirits negotiated with the evil ones until they were willing to let the child go. When the chant reached this point in the story, labor started. Lévi-Strauss showed that the chants "summoned" the woman's body, which means that they reconnected her disordered body with the myth in which the woman had grown up and, in this way, could move her body in the desired direction. Lévi-Strauss emphasized that, to the best of his knowledge, this method was always successful. And most curiously, the shamans carried out their operations intuitively, without really being aware that they were producing their effects through the efficacy of their symbolic framework (the myth).

Twenty-first-century Western man is, in this respect, no different from the Native Brazilians of whom Lévi-Strauss speaks. The Enlightenment man, too, was brought up in a myth, a story that tells something about his origin, that makes him take a certain perspective on life and links his negative and positive emotions and affects to specific stimuli. This myth is the story of the mechanistic universe, the great machine that was set in motion by the Big Bang, in which man is gripped like a small machine in the great machine of the universe. When it comes to sickness and health, then the authority in this story is not the shaman but the medical expert. And that expert, like the shaman, performs a ritual by which he calls the patients' bodies to order. And yes, just like the shaman, the contemporary physician has only a limited awareness of the enormous impact that the symbolic framework within which he operates has on his interventions and he too often believes that psychology has nothing to do with the healings he sees happening in his practice. The enormous contribution of the placebo effect shows us not only how heavily medical practice is based on the impact of visual images, but above all, how overwhelmingly it is based on symbolic effects.

No matter how strong and directly observable the influence of the psychological realm on the physical domain may be, humans—and perhaps Westerners in particular—have a bad habit of focusing attention on the material-biological dimension of life and to consider the psychological realm to be of subordinate importance. And I feel only partially an exception in this regard. However, denying an important determining factor in the causation of a problem usually leads only to an escalation of the problem.

The good news of this story should not go unnoticed, however. The findings on placebos and hypnosis show unequivocally that not only negative images affect the body: Positive images have a similar but inverse effect. I tend to doubt whether we can expect too much from placebos and hypnosis as such. Both have an aspect that is ethically questionable, placebos because they are in essence a form of deception, and hypnosis because the mind of the person hypnotized is subjected to the suggestion of the hypnotist.

Of greater importance are probably the examples of people who, by adhering strictly to ethical principles, have demonstrated they possess the most astonishing physical resilience. In *The Gulag Archipelago*, Solzhenitsyn describes, amongst others, the moving story of Grigory Ivanovich Grigoryev, a prisoner who first spent years in the Nazi concentration camps and then ended up in the gulags under Stalinism. He stood out from everyone else for his legendary honesty and nobility. He refused to carry out assignments that he considered to be unethical, even though he was severely punished for doing so; he refused to participate in the common practice among inmates to steal food from one another when the opportunity presented itself; he adhered strictly to the ethical rules he believed to be appropriate. Solzhenitsyn describes the following about the influence of his spiritual purity on his body:

> *And even more: because of the astounding influence on his body of his bright and spotless human spirit (though no one today believes in any such influence, no one understands it) the organism of Grigory Ivanovich, who was no longer young (close to fifty), grew stronger in camp; his early rheumatism of the joints disappeared completely, and he became particularly healthy after the typhus from which he recovered: in winter he went out in cotton sacks, making holes in them for his head and his arms—and he did not catch a cold.*[21]

One thing is certain: To explore and tap into the possibilities offered by a more psychological approach to human beings, as an alternative to the biological-reductionist approach, is undoubtedly one of the great challenges of the future. If we fail to rise up to this challenge, we are unlikely to find a durable solution to current and future crises.

Our tendency to perceive the above-mentioned scientific observations about psychological causality as strange or unbelievable can only be explained by the fact that ultimately, we are all greatly susceptible to mechanistic-materialistic illusions. But science does not oblige us at all to consider psychological experiences as passively determined by the material domain. On the contrary, the outposts of science—see, for example, the words of Heisenberg, Bohr, Max Planck, and Erwin Schrödinger, as previously cited—came rather to the opposite conclusion. The road to a better understanding of biology and matter will undoubtedly be through the understanding of the structure of our psychological life. For this reason, science must consider as one of its most fundamental tasks to map out the structure of the psychological experience, to clarify its laws, and to study the possibilities this gateway to the human being might open up.

In my opinion, matters such as the placebo effect have to be scientifically investigated. They should not give rise to an immersion in an esotericism that is at odds with the intellect. With his structural anthropology, Lévi-Strauss showed that it is indeed possible to almost entirely describe the effects of stories and images in a rational way. His description is breathtakingly rigorous in scientific terms and at the same time radically anti-mechanistic in nature. This is the way to go: a science that does not allow itself to be blinded by mechanistic ideology but which pushes the rational analysis of reality to the maximum, to the absolute limit of the rationally knowable, to the point where reason transcends itself.

CHAPTER 11

Science and Truth

Totalitarianism is the belief that human intellect can be the guiding principle in life and society. It aims to create a utopian, artificial society led by technocrats or experts who, based on their technical knowledge, will ensure that the machine of society runs flawlessly. In this view, the individual is completely subordinated to the collective, reduced to being a cog in the machine of society (see, for example, Bertrand Russell in *The Impact of Science on Society*).[1]

The ideal of a technocratic society was inherent to the Enlightenment tradition, especially in its positivist branch. Positivist thinkers like Henri de Saint-Simon and Auguste Comte expressed their fanatical belief in a humanistic-technocratic society in which scientists and technocrats would take the place of popes and priests.[2] Not God, but human Reason should be glorified. This is the way to a utopian society without war or conflict, a Realm of Freedom.

Nazism, and even more so Stalinism, are the most ambitious historical attempts to put totalitarian ideology into practice. They would realize paradise, and to this end, everything was considered justified: exclusion, stigmatization, and ultimately industrial extermination of every population group that did not fit within the ideal image. In both historical examples, the new utopian society had to be created through the ruthless application of a rock-solid logic (see chapter 7).

However, it would be a capital mistake to identify the phenomenon of totalitarianism only in totalitarian regimes. There is an ever-present, totalitarian undercurrent that consists of a fanatical attempt to steer and control life in far-reaching ways on the basis of technical, scientific knowledge. Technocratic thinking always walks on two legs. On the one hand, it appeals to people by intimating a positive image of an artificial paradise with which it claims we can be delivered from all adversity and suffering. On the other hand, it imposes itself based on anxiety, as a necessity to solve problems. With every "object of anxiety" that has emerged in our society in recent decades—terrorism, the climate problem, the coronavirus—this process has leapt forward. The threat of terrorism induces the necessity of a surveillance apparatus, and our privacy is now seen as an irresponsible luxury; to control climate problems, we need to move to lab-printed meat, electric cars, and an online society; to protect ourselves against COVID-19, we have to replace our natural immunity with mRNA vaccine–induced artificial immunity.

The fourth industrial revolution, in which man is expected to physically merge with technology—the transhumanist ideal—is increasingly seen as an unavoidable necessity. The entire society has to change into an *internet of bodies*, in which the human body is digitally monitored, tracked, and traced by a technocratic government. This is the only way we will be able to master the problems of the future. There is no alternative. Anyone who refuses to go along with the technological solution is naive and "unscientific."

* * *

Totalitarianism and technocracy like to present themselves as the pinnacle of rationality and science. The technocratic paradise will make the population happy and healthy; or at least offer the greatest chance of achieving this. With subcutaneous sensors, every biochemical change can be registered and reported. Anyone showing signs of illness can be immediately examined and receive adequate treatment. In order to achieve this in an efficient way, everything has to be permanently and monotonously exposed to the artificial light of monitoring and government control. The fact that the human being is like a flower that only blooms when it can enjoy the shade of privacy once in a while is of minor importance in a technocratic worldview. Anyone who refuses to go along with the system lacks civic sense, considers oneself more important than the collective. Your health is no longer your personal business, because some diseases are contagious. However, even within an objectifying biological-reductionist perspective, it has been clear for decades that too much (government) control is harmful to health in itself. To use the example of a viral infection: Control leads to stress and stress in its turn leads to a greatly reduced physical resistance in viral infections (see chapter 10, for example, up to 80 percent more mortality). Acting on the basis of a biological-reductionist analysis is effectively a recipe for failure, even on a purely physical level. One cannot understand the course of a viral infection on the basis of the mechanistic processes seen through the small ring light of a microscope—the whole psychological, sociological, and economic context plays an essential role. Hegel already knew that "Das Wahre ist das Ganze" [The truth is the whole].[3]

This is exactly what twentieth-century science has primarily shown us in an astonishing way: All things small and all things large are connected, everything is part of an overarching, complex, and dynamic system.

In order to understand the course of a viral disease—and more broadly, health and happiness—we have to contemplate man and society and observe the principles of nature. This way, the great questions of life, which were relegated to the second plane by mechanistic ideology, are brought to the fore again: Who are we as desiring beings? How do we relate to other people, to our bodies, to pleasure, to nature, to death? What is our place in nature? There will never be a definitive answer to these questions. Each person has to reformulate the answers to these questions in every new situation, and they can never be definitively determined in a purely rational way (see chapter 9). The end point of science is not reached with a perfectly rational understanding and control of reality; instead, it lies in the final acceptance that there are limits to human rationality, that knowledge does not belong to man, but has to be situated in the wider system of which man forms a part.

* * *

Herewith, we have arrived at an interesting field of tension. On the one hand, you can see the development of science as a steady growth of rational knowledge, as an ever-increasing multitude of phenomena show us which laws they obey. But on the other hand, you can also see the course of science as a process that leads to an a-rational core in things, to something that eludes human understanding. And this *something* is not just a negligibly minor aspect of all things observed, it is the very essence of life (see chapter 3). It's at this level that we can discern that, as the rationalization of the world continues, human beings also increasingly feel that the essence of life is eluding them and that they are more and more often confronted with experiences of meaninglessness, anxiety, psychological discomfort, and frustration (part 1). It is to be expected that the series of crises in which we find ourselves will make the inconsistencies in the mechanistic ideology and the failure of associated pseudo-rational remedies increasingly apparent, and a certain group of people will see more and more clearly what the founders of science already saw: The essence of things is not rationally knowable, and reality cannot be reduced to mechanistic frameworks. When realizing this, we can finally start to look for the essence of life where it truly can be found: in that which always escapes rationalization and mechanization, in that which disappears from a conversation when you digitalize it, in the difference between the mother's womb and an artificial plastic womb, in the difference between the heat of an electric heater and that of a wood-burning stove, and so on.

* * *

The journey of science does not end in superior knowledge but in a kind of Socratic *modesty*. A human who has traveled this journey far enough knows better—he just knows—that all rational knowledge is relative and remains alien to the essence of the object he is trying to understand. At the end of this journey awaits an encounter with something that cannot be grasped with logic and rationality. The great minds of science have testified to such encounter in many different ways. Albert Einstein liked to talk about the elusive mystery that he found everywhere in the universe and about the wonderful structure of reality. Niels Bohr understood that poetry has more grip on all things Real than logic.[4] And Max Planck said that all matter is grounded in a conscious and intelligent Mind that holds the fate of the world and every human being in its almighty hand:

> *As a man who has devoted his whole life to the most clear-headed science, to the study of matter, I can tell you as a result of my research about the atoms this much: There is no matter as such! All matter originates and exists only by virtue of a force which brings the particles of an atom to vibration and holds this most minute solar system of the atom together. . . . We must assume behind this force the existence of a conscious and intelligent Mind. This Mind is the matrix of all matter.*
>
> *Both religion and science require a belief in God. For believers, God is in the beginning, and for physicists He is at the end of all considerations. To the former He is the foundation, to the latter, the crown of the edifice of every generalized world view.*
>
> *That God existed before there were human beings on Earth, that He holds the entire world, believers and non-believers, in His omnipotent hand for eternity, and that He will remain enthroned on a level inaccessible to human comprehension long after the Earth and everything that is on it has gone to ruins; those who profess this faith and who, inspired by it, in veneration and complete confidence, feel secure from the dangers of life under protection of the Almighty, only those may number themselves among the truly religious.*[5]

It is the rule rather than the exception that the founders of science left the rationalistic worldview behind them. Just have a look at their more contemplative works—Einstein, Werner Heisenberg, Erwin Schrödinger, Louis de Broglie, Planck, Bohr, Wolfgang Pauli, Sir Arthur Eddington, Sir James Jeans—all of them had a mystical worldview because they were confronted in their research objects with an irresolvable mystery.[6] That in no way means a minimization of the importance of rationale and logic. But it does mean that rationality is not humanity's final destination. Humanity has to step firmly onto the path of logic in order to ultimately transcend rationality.

* * *

Great scientists have left the logical-factual discourse of science behind and returned in an enlightened way to the type of discourse that during the Enlightenment was initially deemed subordinate: a poetic or mystical discourse, a discourse that shows an original respect and a genuine awe for the unnameable, for that which time and again eludes the human mind. Here, we see something interesting: The trajectory that science took is structurally identical to the trajectory that every human child (or at least the majority of children) takes during the transformation into a subject. I'll repeat the developmental psychological reasoning I presented in chapter 5 in order to put this in a broader perspective.

Each child starts life in a symbiotic resonance with the mother, which is realized through early (body) language. From the mirror stage, this direct resonance comes to an end. From then on, the child stubbornly tries to determine in a logical way which word refers to which object. The ultimate object it tries to get a grip on is always the desire of the Other. What does the Other want? Ultimately, the eagerness to understand the discourse of the Other always arises from the urge to *become* the Other's object of desire. This position, on the one hand, opens up a prospect of narcissistic pleasure and, on the other hand, induces an immersion in dependence and anxiety. The persistent attempts to fixate the meaning of words deprive them of their ability to induce symbiosis; the fixation of their meaning causes the words to lose their resonating power and the sounds no longer produce the connection they produced in the first months of life. This way, we see a connection between a number of elements: fanatic pursuit of logical-rational understanding, narcissism, dependence, anxiety, social isolation.

Around the age of three and a half, after the mirror stage, a second enormous revolution takes place in the subjective experience of the child. It starts to realize that words cannot have a definitive meaning—he comes to realize that human language is affected by an irresolvable lack and that there can never be any definitive certainty. The narcissistic illusion of becoming the ultimate object of the Other's desire is shaken and, at first, the child is inevitably confronted with the primal fear in the narcissistic universe: being left behind as an object for disposal that does not meet the requirements of the Other. At that point, the child can choose between two possible paths. On the first path, it shies away from the narcissistic fear and tries to undo the uncertainty by clinging even more stubbornly to narcissism and (pseudo)rationality. This way, it inevitably slides into an increasingly isolated existence and, ultimately, also into more and more anxiety and unease.

The second possibility is that the child discovers in that uncertainty the space to give substance to life in a creative way and to develop individuality: No longer having to be the object of the Other opens up a space to be oneself and to realize one's own personality. The child no longer aspires to the enjoyment of being the object of the Other but rather to being liked in its individuality as a human being; in its own, personal way in which it makes choices and relates to other people as a human being. On this path, children become increasingly sensitive to nonfactual and nonlogical use of language, a use of language that shows individuality and creativity. It is precisely by practicing this use of language that the child partly rediscovers the resonating function of language and the connection with the Other. The flexibility of such use of language, the fact that not every word has to be linked to one specific meaning, allows the exchange of sounds to transfer something of the (logically elusive) individuality of interlocutors to each other. At this point, speaking changes from a vehicle for transferring knowledge into subjective truth.

On this path, the child will, in all respects, make the transition from the narcissistic position of *his majesty the baby*, from the child who finds it normal that the Other is always there for him, to its position of a human among other humans. In this transformation, it also emancipates itself. It is no longer dependent on the parents to know what is allowed and what is not allowed, what is accepted and what is not accepted in every new situation, and it becomes aware of the broad principles that regulate human relationships and which it has to substantiate itself to a certain extent. Here as well, we can see a connection between a number of elements: ability to tolerate uncertainty, sensitivity to resonating language, humanism, individuality, sovereignty, connection with the Other.

This revolution takes place in different degrees in every child and it is never conclusive. In a sense, all of life consists of an attempt to find space for oneself in the relationship with Others. Some people exert themselves intensively toward this goal, others less so, but no one escapes this existential task in life. The more man advances in this process, the more energy and creative power he will have. The ultimate potential that can be realized on this path is unclear, but the enormous influence of the psychological realm on the body, which we discussed in the previous chapter, shows that its possibilities are extraordinary. It is on this track that the future of humanity lies and not on the mechanistic-transhumanistic track.

* * *

Science, as well as the Enlightenment society based thereon, have now arrived at the same crossroads, as encountered by every child when confronted with the fundamental uncertainty of its existence and of its position in relation with the Other. As a society, we can shy away from anxiety and deny our uncertainty, or we can defy our narcissistic anxiety and accept the uncertainty. The first choice means that we look for the solution in an even more (pseudo)scientific ideology, false rationality, false certainty, and technological control; this way, we end up with even more anxiety, depression, and social isolation. And we will respond to that by trying even more stubbornly to control the uncontrollable, resulting in even more despair. In this book, we have shown that the logical end point of this vicious circle is mass formation and totalitarianism, that is, the radical destruction of all human creativity, individuality, diversity, and every form of social connectedness (except the bond between the individual and the state collective). We can see, in all facets of society, how this process is now evolving toward its limit. For the first time in history, the entire global village has been caught up in the same process of mass formation and the "technologization" and "mechanization" of the world has been scaled up to such an extent that the omnipresent control reaches into the core of intimacy and private life. Therefore, we are experiencing the end point of a cycle, the moment at which a ruling ideology is driven to its ultimate consequence, rears up with its full power for one last time, and thereby shows its powerlessness in a definitive and final way.

When choosing the second path, society defies its anxiety and recognizes that uncertainty is inherent in the human condition and is a necessary condition for the emergence of creativity, individuality, and human connectedness. On this path, society becomes a space in which connectedness and individual differences mutually reinforce one another—as opposed to totalitarian systems in which the collectivity radically encroaches upon the individual liberty of every person and where all diversity disappears and is replaced by a monotonous state identity. The Great Science has preceded us on this path—it followed Reason to its absolute limit, whereupon it opened up a view to a new form of knowing, a new form of connecting with the Other, and to a human existence based on different principles.

The way in which it arrived at this point is structurally the same as the process a young child goes through. Young science, too, starts from a belief that the object being studied can be fully understood by means of logical reasoning. Facts are logical—how could they not be? However, the further the logical analysis of the phenomenon under investigation is carried, the more clearly one sees the emergence of a core that is intrinsically illogical and inaccessible to the human mind. And just like with a child, that moment gives rise to an awareness of the relativity of all logic as well as a heightened sensitivity to forms of language that don't aim to be logically understood but lead to a more direct affinity, to resonance with the object (poetry, mysticism, etc.).

I started this book by stating that the emergence of the mechanistic view of the world and mankind was a revolution at the level of acquiring knowledge about the world. Within a religious worldview, knowledge was revealed to man by God. Therefore, the source of all knowledge lay outside man. Within the mechanistic worldview this all changed: Man situated the source of knowledge within himself. He could come to knowledge himself by observing facts and exploring their mutual connections by means of logical reasoning. But at the end of the journey, science again has to conclude that knowledge lies outside of man (see, for example, the quote from Planck earlier in this chapter).

The ultimate knowledge lies outside of man. It vibrates in all things. And man is able to receive it, by tuning his vibrations, like a string, to the frequency of things. And the more man is able to set aside prejudices and beliefs, the more purely he will vibrate with the things around him and receive new knowledge. This is one possible interpretation of René Thom's thesis that great scientists do not necessarily have an exceptional logical-thinking capacity but rather an extraordinary ability to empathize with the things they study (see chapter 1).[7]

Science is only one of the ways that leads to this empathy. Learning a craft also leads to this ability. The starting point is a logically coherent knowledge of the object to be manufactured and of the artisanal procedure to do so. And as you learn to apply that knowledge in a practical way, you develop a feeling with the tools and the materials, which transcends any logical knowledge. This is precisely what constitutes the essence of a craftsman, a feeling—his affinity with and knowledge of his craft, his craftsmanship—can only be acquired through prolonged and disciplined practice. This is the reason you can't become a craftsman merely by accumulating theoretical knowledge.

Learning an art is also an excellent example. At first, you learn a logical, coherent set of rules and after years of practice, you acquire an affinity that transcends these rules. What's more: The rules ultimately become a ballast and have to be thrown overboard. In Japan, there is a proverb that says that one must protect the rules of an art only long enough to be able to break them. Masaaki Hatsumi, 34th Grandmaster of Togakure school of ninjutsu, said that the techniques of his martial art must be learned to be ultimately forgotten.[8] Letting go of the techniques, after they have been practiced and they have trained and cultivated the body, is more difficult than learning them. But it is crucial. Anyone who still needs to think about techniques on the battlefield will die. The same grandmaster also stated that prolonged practice of martial arts leads to the realization that weapons have a will of their own and that you should never enslave them. Each sword has its own character, wants to move in a certain way; only if you can feel where it wants to go, will it bring you what you expect it to do.

The ability to empathize also plays a role in relation to our own body. Our bodies are in essence foreign to ourselves. It responds to all kinds of stimuli—food, other people, all kinds of situations—and they do so autonomously, without our knowledge or volition. We can learn to feel our body throughout our lives, for example through certain movement-based arts or meditation, by attentively observing the effects of all kinds of factors (nutrition, exercise, etc.) on our body, possibly by repeatedly putting our physical experiences into words during psychoanalytic therapy. Whoever listens to his body and learns to understand its language holds the key to health. The feeling with one's own body is more important than any medicine and also more important than any "objective" rational knowledge of, for instance, healthy food.

In the same way, man also has to come to know himself as a psychological being, as a confluence of subjective experiences, thoughts, feelings, especially as they arise in relations with others. The ability to sense one's own experience and to put it into words and to express it in relation to another is what constitutes the core of our existence as human beings. In line with what I discussed in chapter 3, we exist as human beings when we can give something of our individuality to another through full speech—a kind of speaking in which something of the human being we are vibrates and resonates. It is through the art of full speech—which is the art learned, for example, in psychoanalytic therapy—that we are able to realize a real connection with others and the world around us (without thereby losing ourselves).

It is also through this art that we, as human beings, and more broadly as a culture and society, can relate differently to death. Within a mechanistic and biological-reductionistic view of man, suffering, decay, and death can only be *meaningless*; they cannot be seen as something that has something to say and teach us as human beings. This is perhaps the biggest problem with the Great Mechanistic Narrative: The ultimate master of the sublunary—death—has not been given an acceptable part in it. And he doesn't like that. Banned from the story, he terrifies us and creates frantic responses to every threat, whether terrorism or viruses, that end up being more damaging than the problem itself. It is not so much through the belief in a new Great Narrative that our culture will be able to give death a new place but by cultivating the art of integral speaking and by engendering contact with the object. The connection with the Other and the world, the resonance with the wider whole, removes the narrow constraints of the Ego. Literally: To the degree that we can connect with what is outside ourselves, we are able to transcend our own boundaries and our own world of experience gets expanded to an existence that extends endlessly in time and space. Through resonance with the greater plain, we participate in the timelessness of the universe, like a reed rustling in the eternal air of life.

* * *

At the heart of things, there is something that never can be definitively captured in the categories of logic and, therefore, has to be reworded time and again. Each attempt to put it into words can be only ephemeral; each new encounter brings forth new words, words directly born from contact with the object. "Le vrai est toujours neuf" [The truth is always new], said Max Jacob.[9] The encounter with the object produces *truth*, an ever-renewing way of speaking, the core characteristic of which is not so much that it is logically correct but that it resonates freshly and sincerely with what it is about. Poetry, sometimes nonsensical from a logical point of view, can carry a lot more truth than a discourse built up strictly from syllogisms.

Truth has become an anachronistic concept—it sounds old-fashioned. In *The Courage of Truth*, the French philosopher Michel

Foucault makes an interesting distinction between rhetoric and truth.[10] A person who uses rhetoric tries to arouse in another ideas and beliefs that he does not share himself. For someone who adheres to speaking the truth, the reverse is true. He sincerely tries to convey an idea or experience that lives within himself to the Other through his speaking; he tries to make something he feels in himself resonate in an Other.

In recent centuries, and especially in recent decades, the public sphere has been increasingly filled with rhetoric. We were already used to such rhetoric from politicians. No one expected them to even try to fulfill their election promises during their term of office. In the long run, the population simply accepted it: A politician's election discourse only serves to *convince*. And in fact, the same goes for commercials. Only an idiot believes that they paint an accurate picture of the product being advertised. Moreover, during the coronavirus crisis, we learned that it is not really different for those who present themselves as scientists. What they say today is guaranteed to be retracted tomorrow.

The real volte-face and revolution that society has to face is to shake off rhetoric and resolutely turn to truth as a guiding principle. Foucault distinguished four forms of truth-telling: prophecy, wisdom, *techné*, and *parrhesia* (speaking boldly).[11] Each of the four has to do with the ability to resonate with an object and to make that resonance resound in sincere speaking and to transfer it to others. Prophecy is a predictive power that does not come from logical understanding, but—as the great French mathematician and philosopher of science Henri Poincaré suggested—from the ability to sense the story that grips reality. Wisdom is the ability to keep silent and allow the Other to hear his own words. The *techné* is the ability to speak technically correctly, to produce a logical-factual discourse that adequately reflects the structure of the object to which it refers. And finally, the *parrhesia* refers to the courage to publicly express words that break through the fallacious discourse of society. The reappraisal of the phenomenon of truth-telling will be the indicator par excellence of the progress of the revolution, which is necessary to overcome the tendency toward totalitarianism inherent in the Enlightenment tradition.

* * *

Finally, we can ask ourselves: Isn't it dangerous to give up rationality as an ideal? This question prompts me to a small reflection, which only due to the seriousness of its subject is not banal. Thirty-five thousand children die of hunger every day. Why doesn't this upset the masses, while a virus does? In our rational view of humanity, why don't we save these young, hungering lives at a much lower cost than those threatened by the coronavirus, without the risk of losing civil liberties, and without the dangers associated with experimental medical interventions? No one panics for a child that is dying on the other side of the world. This is the inconvenient truth. The rationality and humanism of the Enlightenment are in many ways a masquerade and a fig leaf. Strip man of this masquerade and you look into the eyes of irrationality; look behind the fig leaf of rationality and you will find the ancient human vices.

A rational worldview does not prevent us from giving free rein to irrational thinking. On the contrary, it prevents us from *recognizing* irrationality. And as such, irrationality takes on grotesque proportions. On the other hand, one who knows the limits of his intellect usually becomes less arrogant and more humane, more capable of allowing the other to be different. When his intellect stops shouting, he is able to hear the things of life murmur their own story. He realizes that he is also entitled to his own story. The awareness that no logic is absolute is the prerequisite for mental freedom. The gap in the logic literally opens up a space for our own style and for the desire to create. "I became healthy while creating"—this is how Goethe described his medicine against the ailment that is life. Perhaps, it might also work against viruses?

In any case, this remedy ensures that we can honor the right to free speech and the right to self-determination without feeling threatened by one another. It encompasses the potential to mitigate anxiety, discomfort, frustration, and aggression, without the need for an enemy. This is the point at which we no longer need to lose ourselves in the crowd to experience meaning and connectedness, this is the point where the winter of totalitarianism gives way to a new spring of life.

ACKNOWLEDGMENTS

We cannot describe in words where words come from. But we do know where words go—they are always on their way to Another. Man is a narrow passage through which words pass on their journey from source to Other.

The words that found their place in this book rested for years in scribbles and notes. It was the coronavirus crisis that finally moved me to send them out into the world. It was during the crisis that an Other arose for whom those words longed. I would like to thank the people who were open to what I had to say in opinion pieces, podcasts, and interviews. It was their human responses—as I received them through social media, emails or letters—that allowed the words to blossom within me and that gave me the desire to continue to speak and to put my thoughts down in writing.

I would like to thank the many people who offered me a forum from which to speak out. I am thinking in particular of Marlies Dekkers and Ad Verbrugge. Meanwhile, the studio of *De Nieuwe Wereld* feels familiar—to get together with a glass of wine after the recordings feels like coming home. The circumstances under which this book was written made speaking and writing a delicate matter—an act that had to be performed amid great social resistance. I would like to thank the people who shared the experience of going against this resistance and, in doing so, came into my life in an unexpected way, only to become dear friends. You are too numerous to list here, but you all know, every single one of you, that it is with you in mind that I write these words. You will forever hold a special place in my heart and in my thoughts.

In August 2021, I effectively took up the pen and started writing the text of this book. I was strictly held to a predetermined writing schedule by publisher Nancy Derboven—thank you for this! In the same vein, I want to thank Margo Baldwin for her enthusiasm about my book. Els Vanbrabant and Brianne Goodspeed: Thank you so much for your unwavering commitment to deliver a high-quality English translation of this book. Special thanks go to Dr. Robert Malone for his continuing efforts to bring my work to the attention of the Anglo-Saxon world—Robert, it was great meeting you in Spain and I hope we will meet again in the future!

I collected the many thoughts and musings on totalitarianism from my scientific journal, opinion pieces, and articles and let them merge into the text of this book. I would like to thank the people who have read and commented on the draft chapters of this book during the writing process: Debora Desmet, Liesje Breyne, Nathalie De Neef, Steven Wouters, and Tineke De Cock. Without your willingness to be my sounding board, the text-to-be would never have matured. Debora, my youngest sister, thank you for making me reconsider verb tenses and reminding me time and again of Nietzsche's ten writing commandments; Liesje, you always found simpler forms of expression where words were entangled in stiff knots; Nathalie, your amusing comments and suggestions of temperance and moderation kept me on the right track; Steven, thank you for providing me with crucial additional references and corrections; Tineke, thank you so much for taking the text to the next level by critically questioning every sentence, up to the last letter, and mercilessly demanding logical clarity. And finally: Valerie—thank you for proofreading, but above all, for putting up with my absentmindedness and short nights during the months leading up to the birth of this book and for always being there to listen to my endless reflections and thought improvisations.

MATTIAS DESMET, November 2021, Meigem

NOTES

Introduction

1. Hannah Arendt, *The Origins of Totalitarianism* (London: Penguin Books, 1951): 622.
2. Maaike Schwering, "Himalaya voor het eerst in dertig jaar zichtbaar door schonere lucht" [Himalayas visible for the first time in thirty years through cleaner air], *Knack*, August 4, 2020, https://weekend.knack.be/lifestyle/reizen/natuur/himalaya-voor-het-eerst-in-dertig-jaar-zichtbaar-door-schonere-lucht/article-news-1586287.html.

Chapter 1: Science and Ideology

1. Immanuel Kant, "Beantwortung to the Frage: Was it Aufklärung?" [Answer to the question: What is Enlightenment?], *Berlinische Monatsschrift* (December 1784): 481–94.
2. Michel Foucault, *De moed tot waarheid* [The courage to truth] (Amsterdam: Boom, 1978).
3. Max Jacob, *Cornet a dés* [Dice box] (Paris: Jourde and Allard, 1917).
4. W. Heisenberg, "Über den anschaulichen Inhalt der quantentheoretischen Kinematik und Mechanik" [On the physical content of the quantum theoretical kinematics and mechanics], *Zeitschrift für Physik* 43, (1927): 172–98.
5. René Thom, *Prédire n'est pas expliquer* [To predict is not to explain], Champs sciences, Editions Eshel, trans. Roy Lisker (IHES edition, 2010): 92.
6. Elisabeth Margaretha Bik, Arturo Casadevall, and Ferris Fang, "The Prevalence of Inappropriate Image Duplication in Biomedical Research Publications," *mBio* 7, no. 3 (July 2016): e00809-16.

7. Owen Jarus, "Famed Archaeologist 'Discovered' His Own Fakes at 9000-Year-Old Settlement," *Live Science*, March 12, 2018, https://www.livescience.com/61989-famed-archaeologist-created-fakes.html.
8. I. M. D. Souza and A. M. L. Caitite, "The Amazing Story of the Fraudulently Cloned Embryos and What It Tells Us about Science, Technology, and the Media," *Historia, Ciencias, Saude—Manguinhos* 17, no. 2 (2009): 471–93.
9. Joseph Hixson, *The Patchwork Mouse* (Garden City, New York: Anchor Press, 1976).
10. Isabelle De Groote et al., "New Genetic and Morphological Evidence Suggests a Single Hoaxer Created 'Piltdown Man'," *Royal Society of Open Science* 3, no. 8 (August 2016): 160328, https://doi.org/10.1098/rsos.160328.
11. Gretchen Vogel, "Psychologist Accused of Fraud on 'Astonishing Scale'," *Science* 334, no. 6056 (November 4, 2011): 579–79, https://doi.org/10.1126/science.334.6056.579.
12. Daniele Fanelli, "How Many Scientists Fabricate and Falsify Research? A Systematic Review and Meta-analysis of Survey Data," *Plos One* 4, no. 5 (2009): e5738, https://doi.org/10.1371/journal.pone.0005738.
13. Mona Baker and Dan Penny, "Is There a Reproducibility Crisis?" *Nature* 533 (May 26, 2016): 452–54.
14. C. Glenn Begley and Lee M. Ellis, "Drug Development: Raise Standards for Preclinical Cancer Research," *Nature* 483 (March 2012): 531–33, https://doi.org/10.1038/483531a.
15. Andrew Chang and Phillip Li, "Is Economics Research Replicable? Sixty Published Papers from Thirteen Journals Say 'Usually Not'," Finance and Economics Discussion Series 2015-083 (September 2015): http://dx.doi.org/10.17016/FEDS.2015.083, retrieved from https://www.federalreserve.gov/econresdata/feds/2015/files/2015083pap.pdf.
16. C. Glenn Begley and John P. Ioannidis, "Reproducibility in Science: Improving the Standard for Basic and Preclinical Research," *Circulation Research* 116, no. 1 (January 2015): 116–26, https://doi.org/10.1161/CIRCRESAHA.114.303819.

17. John P. Ioannidis, "Why Most Published Research Findings Are False," *PLoS Medicine* 2 (August 2005): e124, https://doi.org/10.1371/journal.pmed.0020124.
18. Mattias Desmet, *The Pursuit of Objectivity in Psychology* (Ghent: Borgerhoff & Lamberigts, 2018).
19. G. J. Meyer et al., "Psychological Testing and Psychological Assessment: A Review of Evidence and Issues," *American Psychologist* 56, no. 2 (February 2001): 128–65.

Chapter 2: Science and Its Practical Applications

1. Benjamin Kidd, *The Science of Power* (New York/London: Putnam's Sons, 1918): 18–19.
2. David Graeber, *Bullshit Jobs* (Amsterdam: Business Contact, 2018): 16.
3. Graeber, *Bullshit Jobs*: 23.
4. Graeber, *Bullshit Jobs*: 27.
5. Graeber, *Bullshit Jobs*: 18.
6. R. M. Giusti, K. Iwamoto, and E. E. Hatch, "Diethylstilbestrol Revisited: A Review of the Long-Term Health Effects," *Annals of Internal Medicine* 122, no. 10 (May 1995): 778–88, https://doi.org/10.7326/0003-4819-122-10-199505150-00008.
7. Arthur Shapiro, *The Powerful Placebo: From Ancient Priest to Modern Physician* (Baltimore: The Johns Hopkins University Press, 1997).
8. Bruce Wampold et al., "The Placebo Is Powerful: Estimating Placebo Effects in Psychotherapy and Medicine from Randomized Clinical Trials," *Journal of Clinical Psychology* 61, no. 7 (July 2005): 835–54, https://doi.org/ 10.1002/jclp.20129.
9. Gaia, "Nieuwe cijfers: wereldwijd 79,9 miljoen dierproeven" [New figures: 79.9 million animal tests worldwide], *Gaia*, April 24, 2020, https://www.gaia.be/nl/nieuws/wereldproefdierendag-nieuwe-cijfers-wereldwijd-799-miljoen-dierproeven.

Chapter 3: The Artificial Society

1. James Gleick, *Chaos: Making a New Science* (London: Penguin Books, 1987): 292.
2. Gleick, *Chaos*: 41–44.

3. Gleick, *Chaos*: 43.
4. Mattias Desmet, "Waarom digitale gesprekken zo uitputtend zijn" [Why digital conversations are so exhausting], *Knack*, May 6, 2020, https://www.knack.be/nieuws/wetenschap/waarom-digitale-gesprekken-zo-uitputtend-zijn/article-opinion-1606309.html.
5. A. Hautekeet, "Online leven is schadelijker dan coronavirus" [Life online is more harmful than the coronavirus], *De Standaard*, May 26, 2020, https://www.standaard.be/cnt/dmf20200525_04971253.
6. C. De Kock, "Om echt te kletsen moet je kunnen klinken" [You can't have a good chat without raising a glass together], *De Standaard*, May 27, 2020.
7. P. Cabenda, "Met slimme seksspeeltjes kun je—heel veilig—van elkaar genieten" [With smart sex toys you can have sex without risk], *De Morgen*, May 26, 2020.
8. JCA, "Maleisiër via Zoom ter dood veroordeeld" [Malaysian sentenced to death via Zoom], *De Standaard*, May 20, 2020, https://www.standaard.be/cnt/dmf20200520_04966951.
9. S. Kelepouris, "'De slinger van het thuiswerken is doorgeslagen': experte ergonomie Veerle Hermans" ["The pendulum of working from home has swung": ergonomics expert Veerle Hermans], *De Morgen*, May 21, 2020, https://www.demorgen.be/nieuws/de-slinger-van-het-thuiswerken-is-doorgeslagen-experte-ergonomie-veerle-hermans~bc2fcac7.
10. Patricia Kuhl, "Is Speech Learning 'Gated' by the Social Brain?" *Developmental Science* 10, no. 1 (January 2007): 110–20, https://doi.org/10.1111/j.1467-7687.2007.00572.x.
11. Annie Murphy-Paul, *Origins: How the Nine Months Before Birth Shape the Rest of Our Lives* (Amsterdam: Free Press, 2011).
12. G. di Pellegrino et al. "Understanding Motor Events: A Neurophysiological Study," *Experimental Brain Research* 91, no. 1 (1992): 176–80.
13. Gianpiero Petriglieri, "I spoke to an old therapist friend today, and finally understood why everyone's so exhausted after the video calls. It's the plausible deniability of each . . ." Twitter, 4:43 PM, April 3, 2020, https://twitter.com/gpetriglieri/status/1246221849018720256.

14. Mattias Desmet, "Some Preliminary Notes on Empirical Test of Freud's Theory on Depression," *Frontiers in Psychology* 4 (May 2013): 158, http://dx.doi.org/10.3389/fpsyg.2013.00158.
15. Marjolijn Vanslembrouck, "Nooit meer die tijd van de maand: volgens artsen is menstrueren 'compleet nutteloos'" [Never that time of the month again: according to these doctors, menstruating is "completely useless"], *De Morgen*, July 19, 2020, https://www.hln.be/fit-en-gezond/nooit-meer-die-tijd-van-de-maand-volgens-artsen-is-menstrueren-compleet-nutteloos~ac76aebc.
16. Emily A. Partridge et al., "An Extra-Uterine System to Physiologically Support the Extreme Premature Lamb," *Nature Communications* 8: 15112, https://doi.org/10.1038/ncomms15112.
17. Tech Insider, "Concept Incubator Would Grow Your Babies at Home," YouTube, 1:46, July 4, 2017, https://www.youtube.com/watch?v=cgmdF9l7K9o.
18. Tech Insider, "Concept Incubator."
19. PVZ, "Volgens Elon Musk hebben we binned 5 jaar geen menselijke taal meer nodig" [According to Elon Musk, we will no longer need human language within 5 years], *De Morgen*, May 9, 2020, https://www.hln.be/ihln/volgens-elon-musk-hebben-we-binnen-5-jaar-geen-menselijke-taal-meer-nodig~a35bc439.
20. Sven de Jong, "Geo-engineering als laatste redmiddel" [Geoengineering as a last resort], *Nemo Kennislink*, January 8, 2010, https://www.nemokennislink.nl/publicaties/geo-engineering-als-laatste-redmiddel.
21. George van Hal, "Wetenschappers binden strijd aan met anti-aanbaklaag en waterafstotende regenjas" [Scientists challenge the adverse effects of non-stick coatings and water-repellent raincoats] *De Morgen*, July 1, 2020, https://www.demorgen.be/tech-wetenschap/wetenschappers-binden-strijd-aan-met-anti-aanbaklaag-en-waterafstotende-regenjas~b79d1f20.
22. Wim Schepens and Tijs Neirynck, "Supermarkten halen opnieuw tientallen producten uit winkelrekken door ethyleenoxide, wat is er aan de hand?" [Supermarkets once again have to remove dozens of products from store shelves due to ethylene oxide, what's going

on?], *VRT NWS*, September 22, 2021, https://www.vrt.be/vrtnws/nl/2021/08/06/ ethyleenoxide-terugroepingsactie.
23. M. Martin, "Waarschuwing voor directe link tussen chemicaliën en de 'wildgroei' aan beschavingsziekten: zo zit het" [Warning for direct link between chemicals and the "proliferation" of civilization diseases: this is how it is], *De Morgen*, August 7, 2019, https://www.demorgen.be/nieuws/waarschuwing-voor-directe-link-tussen-chemicalien-en-de-wildgroei-aan-beschavingsziekten-zo-zit-het~bd2839f3.
24. Max Weber, "Wissenshaft als Beruf" [Science as a profession], 1919, https://de.wikisource.org/wiki/Wissenschaft_als_Beruf.
25. Hannah Arendt, *The Origins of Totalitarianism* (London: Penguin Books, 1951): 585.
26. Arendt, *The Origins of Totalitarianism*: 507.
27. Plato, *De ideale staa* [The ideal state], *Politeia* (Amsterdam: Athenaeum—Polak & Van Gennep, 2010): 182.
28. Eric Voegelin, "The Origins of Scientism," *Social Research: An International Quarterly* 15, no. 4 (December 1948): 462–94, cited in Arendt, *The Origins of Totalitarianism*: 453.

Chapter 4: The (Im)measurable Universe

1. Benoit Mandelbrot, "How Long Is the Coast of Britain? Statistical Self-Similarity and Fractal Dimensions," *Science* 156, no. 3775 (May 1967): 636–38, https://doi.org/10.1126%2Fscience.156.3775.636.
2. E. H. Simpson, "The Interpretation of Interaction in Contingency Tables," *Journal of the Royal Statistical Society, Series B* 13, no. 2 (July 1951): 238–41, https://doi.org/10.1111/j.2517-6161.1951.tb00088.x.
3. C. Peeters et al., "De PCR test is onbetrouwbaar en het testbeleid faalt" [The PCR test is unreliable and the test policy is failing], September 27, 2020, *HP/De Tijd*.
4. Luc Gochel, "Le Liégeois qui a fait plier les experts" [The Liège professor who causes problems for the experts.] *Sudinfo Lameuse*, December 8, 2020, https://lameuse.sudinfo.be/619193/article/2020-08-12/le-liegeois-qui-fait-plier-les-experts.

5. Scottish Government, "Counting People in Hospitals with COVID-19," September 15, 2020, https://blogs.gov.scot/statistics/2020/09/15/counting-people-in-hospital-with-covid-19.
6. Jeroen Bossaert, "Sjoemelen ziekenhuizen met coronacijfers? Documenten wijzen op 'financiële optimalisatie'" [Are hospitals cheating with corona figures? Documents point to "financial optimization"], *HLN* May 14, 2021, https://www.hln.be/binnenland/hln-onderzoek-sjoemelen-ziekenhuizen-met-coronabeelden-schrijven-wijs-op-financiele-optimisatie~aef03627.
7. Peter C. Gøtzsche, *Dodelijke medicijnen en georganiseerde misdaad: achter de schermen van de farmaceutische industrie* [Deadly drugs and organized crime: behind the scenes of the pharmaceutical industry] (Rotterdam: Lemniscate, 2021).
8. National Center for Health Statistics, "Weekly Updates by Select Demographic and Geographic Characteristics: Provisional Death Counts for Coronavirus Disease (COVID-26)," Centers for Disease Control and Prevention, August 26, 2020, https://stacks.cdc.gov/view/cdc/92550.
9. Redactie, "Studie: '90% coronadoden valt in landen met veel obesitas'" [Study: "90% of corona deaths occur in countries with a lot of obesity"], *De Morgen*, March 6, 2021, https://www.demorgen.be/nieuws/studie-90-coronadoden-valt-in-landen-met-veel-obesitas~ba1823fc.
10. Liaoyi Lin et al., "CT Manifestations of Coronavirus Disease (COVID-19) Pneumonia and Influenza Virus Pneumonia: A Comparative Study," *American Journal of Roentgenology* 216, no. 1 (January 2021): 71–79.
11. E. Ooms, "Wetenschap in haar blote kont: de illusie van de zekerheid der cijfers" [Bare-assed science: the illusion of the certainty of numbers], March 24, 2021, https://www.artsenvoorvrij.be/blog/2021/03/24/wetenschap-in-haar-blote-kont-de-illusie-van-de-zekerheid-der-cijfers-de-speurtocht-naar- sciensanos-108-000-verwachte-doden.
12. Luc Bonneux, "De slechtst georganiseerde ouderenzorg van de EU" [The worst organized elderly care in the EU], *De Standaard*, June 12, 2020, https://www.standaard.be/cnt/dmf20200611_04988858.

13. JVH, "Duitse longarts: 'Grootste fout tijdens eerste golf was massale intubatie'" [German pulmonologist: "Biggest mistake during first wave was massive intubation"], *Het Nieuwsblad*, December 24, 2020, https://www.nieuwsblad.be/cnt/dmf20201224_96693948.
14. Mattias Desmet, "De angst voor het coronavirus is gevaarlijker dan het virus zelf" [The fear of the coronavirus is more dangerous than the virus itself], *VRT NWS*, March 25, 2020, https://www.vrt.be/vrtnws/nl/2020/03/25/angst-voor-het-virus.
15. S. V. Subramanian and Akhil Kumar, "Increases in COVID-19 Are Unrelated to Levels of Vaccination across 68 Countries and 2947 Counties in the United States," *European Journal of Epidemiology*, 36 (2021): 1237–40, https://doi.org/10.1007/s10654-021-00808-7.
16. A. R. Brock, "Spontaneous Abortions and Policies on COVID-19 mRNA Vaccine Use During Pregnancy," *Science, Public Health Policy, and the Law* 2021, no. 4 (2021): 130–43.
17. Günter Kampf and Martin Kulldorf, "Calling for Benefit-Risk Evaluations of COVID-19 Control Measures," *The Lancet* 397, no. 10274 (February 13, 2021): 576–77, https://doi.org/10.1016/S0140-6736(21)00193-8.
18. Oxfam, "Honger kan eind 2020 dodelijker worden dan coronavirus zelf" [Hunger could become more deadly than the coronavirus itself by the end of 2020], *Knack*, June 9, 2020, https://www.knack.be/nieuws/wereld/oxfam-honger-kan-eind-2020-dodelijker-worden-dan-coronavirus-zelf/article-news-1618511.html; World Health Organization, "Impact of COVID-19 on People's Livelihoods, Their Health and Our Food Systems," October 13, 2020, https://www.who.int/news/item/13-10-2020-impact-of-covid-19-on-people's-livelihoods-their-health-and-our-food-systems; and United Nations World Food Program, "2020—Global Report on Food Crises," April 20, 2020, https://www.wfp.org/publications/2020-global-report-food-crises.
19. House of Commons, "Science and Technology Committee—Oral Evidence: UK Science, Research and Technology Capability and Influence in Global Disease Outbreaks, HC 136," April 16, 2020, https://committees.parliament.uk/download/file

/?url=%2Foralevidence%2F289%2Fdocuments%2F3825%3Fconvertiblefileformat%3Dpdf&slug=oe00000289pdf.
20. Jon Miltimore, "How Finland and Norway Proved Sweden's Approach to COVID-19 Works," *FEE Stories*, November 13, 2020, https://fee.org/articles/how-finland-and-norway-proved-sweden-s-approach-to-covid-19-works.
21. Hannah Arendt, *The Origins of Totalitarianism* (London: Penguin Books, 1951): xxxviii.
22. Arendt, *The Origins of Totalitarianism*: 622.
23. Arendt, *The Origins of Totalitarianism*: 621.

Chapter 5: The Desire for a Master

1. Sofie Geusensand Jens Vancaeneghem, "Bromfiets gevaarlijkst voor schoolverkeer: 'Puber op brommer naar school in de spits is vragen om problemen'" [Mopeds most dangerous for school traffic: "Adolescents on a moped to school in rush hour is asking for problems"], *De Standaard*, December 20, 2020, https://www.standaard.be/cnt/dmf20200212_04845116.
2. MTM, "Een verfrissende duik in de rivier of de vijver de komende dagen? Geen goed idee: 'Te gevaarlijk, doe het niet'" [A refreshing dip in the river or pond in the coming days? Not a good idea: "Too dangerous, don't do it"], *Het Nieuwsblad*, June 23, 2019, https://www.nieuwsblad.be/cnt/dmf20190623_04475442.
3. BELGA, "Orale seks veroorzaakt meer keelkanker bij Belgen" [Oral sex causes more throat cancer in Belgians], *VRT NWS*, May 8, 2014, https://www.vrt.be/vrtnws/nl/2014/05/08/orale_seks_veroorzaaktmeerkeelkankerbijbelgen-1-1961069.
4. Marjan Temmerman, "Is elkaar de hand schudden voorgoed verleden tijd? 'Ik hoop het een beetje,' zegt viroloog Marv Van Ranst" [Is shaking hands a thing of the past? "I kind of hope so," says virologist Marc Van Ranst], *VRT NWS*, June 16, 2020, https://www.vrt.be/vrtnws/nl/2020/06/16/elkaar-de-hand-schudden-voorgoed-verleden-tijd.
5. EVDG, "Zelfs naast een roker zitten die niet rookt, kan schadelijk zijn voor de gezondheid" [Even sitting next to a smoker who is

not smoking can be harmful to health], *De Standaard*, July 7, 2020, https://www.standaard.be/cnt/dmf20200306_04879315.

6. Henry David Thoreau, *Walden: Life in the Woods* (Utah: Gibbs M. Smith Inc, 2017): 77.
7. Annick Wellens, "Onweersschade? Verzekeringsmaatschappij legt uit wat je moet doen" [Lightening and thunderstorm damage? Your insurance company will explain what to do], *HLN*, September 8, 2018, https://www.hln.be/binnenland/onweersschade-verzekeringsmaatschappij-legt-uit-wat-je-moet-doen~a61148aa.
8. Donna van der Kolk, "Sterren verzekeren hun benen, kont en . . . sperma" [Stars insure their legs, ass and . . . sperm], *Metro*, March 10, 2015, https://www.metronieuws.nl/entertainment/2015/03/sterren-verzekeren-hun-benen-kont-en-sperma.
9. Dylan Haegens, "10 gekste verzekeringen!" [10 craziest insurance policies!], September 14, 2014, YouTube video, 2:33, https://www.youtube.com/watch?v=KNpc6jHjXQA.
10. Koen Snoekx, "Proffen stellen screening borstkanker in vraag: 'Niet minder kans om kankerpatiënt te worden, integendeel'" [Professors question breast cancer screening: "No less chance of becoming a cancer patient, on the contrary"], *De Standaard* March 16, 2019, https://www.standaard.be/cnt/dmf20190316_04261562.
11. Michaéla Schippers, "For the Greater Good? The Devastating Ripple Effects of the COVID-19 Crisis," *Frontiers in Psychology* 11 (September 29, 2020): https://doi.org/10.3389/fpsyg.2020.577740.
12. Michel Foucault, *Histoire de la folie à l'âge classique* [A history of insanity in the age of reason] (Paris: Gallimard, 1972).
13. Manon Dupont, "Universiteit Gent leert studenten 'legaal flirten': 'Ik heb nog geen klacht gehad, dus ik flirt wel goed'" [Ghent University teaches students about "legal flirting": "I haven't had a complaint yet, so I assume my flirting is OK"], *VRT NWS*, January 23, 2020, https://www.vrt.be/vrtnws/nl/2020/01/23/universiteit-gent-leert-studenten-legaal-flirten-ik-heb-nog.
14. Michiel Martin, "Studentendopen liggen meer dan ooit onder vuur: 'Meisjes kregen een banaan om te tonen wat ze ermee konden. Vreselijk'" [Freshmen baptisms are under fire more than ever

before: "Girls were given a banana to show what they could do with it. Despicable"], *De Morgen*, September 15, 2021, https://www.demorgen.be/nieuws/studentendopen-liggen-meer-dan-ooit-onder-vuur-meisjes-kregen-een-banaan-om-te-tonen-wat-ze-ermee-konden-vreselijk~b1e908633.

15. A. G. Fransen, "Pas als het sekscontract getekend is, mogen de Zweden vrijen" [Only when the sex contract has been signed, the Swedes are allowed to have sex] *De Morgen*, June 20, 2018, https://www.demorgen.be/nieuws/pas-als-het-sekscontract-getekend-is-mogen-de-zweden-vrijen~b5b55230.
16. P. van Tyghem, "Zijn blote borsten gevaarlijker dan de Holocaust ontkennen?" [Are bare breasts more dangerous than denying the Holocaust?], *De Standaard*, July 23, 2018, https://www.standaard.be/cnt/dmf20180722_03627256.
17. MVO, "Netflix legt vreemde regels op tegen misbruik: 'Iemand niet langer dan 5 seconden aankijken'" [Netflix imposes strange rules against abuse: "Don't look at someone for more than 5 seconds"], *HLN*, June 14, 2018, https://www.hln.be/showbizz/netflix-legt-vreemde-regels-op-tegen-misbruik-iemand-niet-langer-dan-5-seconden-aankijken~af8736b6.
18. M. Boudry, "Ook links omarmt ontkenners" [The left also embraces deniers], *De Standaard*, May 11, 2019, https://www.standaard.be/cnt/dmf20190510_04390639.
19. Wim Winckelmans, "Nieuwe coronaregels eindelijk bekend: openingsdans kan, polonaise liever niet" [New coronavirus rules finally known: opening dance is possible, polonaise rather not], *De Standaard*, June 30, 2020, https://www.standaard.be/cnt/dmf20200630_94008414.
20. Stijn Cools, "Online zingt het vogeltje ranziger dan ooit tevoren" [Online, the bird is singing more raunchy than ever before], *De Standaard*, June 17, 2020, https://www.standaard.be/cnt/dmf20200616_04992948.
21. Doha Madani, "JK Rowling Accused of Transphobia after Mocking 'People who Menstruate' Headline," *NBC News*, June 7, 2020, https://www.nbcnews.com/feature/nbc-out/j-k-rowling-accused-transphobia-after-mocking-people-who-menstruate-n1227071.

22. TTR, "Duitse verzekeraars willen alcoholslot in alle nieuwe auto's in Europese Unie" [German insurers want alcohol locks in all new cars in the European Union], *HLN*, January 26, 2020, https://www.hln.be/buitenland/duitse-verzekeraars-willen-alcoholslot-in-alle-nieuwe-auto-s-in-europese-unie~a12f69c7.
23. Michael Persson, "Opiniechef *New York Times* sneuvelt na rechts opruiend stuk" [*New York Times* editor in chief has to resign after right-wing incendiary opinion piece] *De Morgen*, June 11, 2020, https://www.demorgen.be/politiek/opiniechef-new-york-times-sneuvelt-na-rechts-opruiend-stuk~b1ea2933.
24. TIB, "'Fawlty Towers' te racistisch voor BBC" ["Fawlty Towers" is too racist for BBC], *Het Nieuwsblad*, January 26, 2013, https://www.nieuwsblad.be/cnt/dmf20130126_022; BELGA, "San Francisco verwijdert standbeeld Columbus" [San Francisco removes Columbus statue], *De Morgen*, June 19, 2020, https://www.zeelandnet.nl/nieuws/san-francisco-verwijdert-standbeeld-columbus; and GHO, "Met corona besmette man slaat op de vlucht in Australië, polite opent jacht op 'volksvijand nummer één'" [Man infected with coronavirus flees in Australia, police open a man hunt for "public enemy number one"], *Gazet van Antwerpen*, August 24, 2021.
25. Sigmund Freud, *Cultuur en Religie 4: Totem en taboe* [Culture and religion 4: Totem and taboo] (Amsterdam: Boom, 1984): 39.
26. James Frazer, *Totemism and exogamy* (London: McMillan, 1910): retrieved from https://archive.org/details/totemismexogamyt01fraz.
27. David Graeber, *Bullshit Jobs* (Amsterdam: Business Contact, 2018): 155.
28. Graeber, *Bullshit Jobs*: 170.
29. Graeber, *Bullshit Jobs*: 49.
30. Erich Fromm, *Escape from Freedom* (New York: Rinehart & Co., 1952).
31. James Anthony, "Australian Government Plans Chinese-Style 'Social Credit' System for Social Media Users," *The Post Millennial*, September 2, 2021, https://thepostmillennial.com/watch-australian-government-plans-chinese-style-social-credit-system-for-social-media-users.

32. Mathieu Verstichel, "Sint-Niklaas heeft vanaf 2022 een eigen digitale stadsmunt: '125 handelaars doen al mee'" [The town of Sint-Niklaas will have its own digital city currency from 2022: "125 retailers are already participating"], *VRT NWS*, June 15, 2021, https://www.vrt.be/vrtnws/nl/2021/06/15/betalen-mensen-hun-brood-in-2022-met-een-digitale-stadsmunt-in-s.
33. R. Andersen, "Opgepakt door het algoritme: hoe China met orwelliaanse technologie massaal burgers vastzet" [Picked up by the algorithm: how China is detaining civilians en masse with Orwellian technology], *De Morgen*, November 26, 2019, https://www.demorgen.be/politiek/opgepakt-door-het-algoritme-hoe-china-met-orwelliaanse-technologie-massaal-burgers-vastzet~b1b6aa682.
34. Tobias Santens, Gianni Paelinck, "Peeters en De Block: 'Dit is alarmerend. Label voor rundvlees moet onderzocht worden'" [Peeters and De Block: "This is alarming. Beef label needs to be investigated"], *VRT NWS*, November 16, 2018, https://www.vrt.be/vrtnws/nl/2018/11/16/peeters-over-onbetrouwbaar-kwaliteitslabel-rundsvlees-zeer-ala.
35. Isabelle Saporte, *Vino Business: The Cloudy World of French Wine* (New York: Grove Press, 2016).
36. MV, VHN, and LOB, "Rubicon, de geheime 'inlichtingencoup van de eeuw'" [Rubicon, the secret "intelligence coup of the century"], *De Standaard*, February 12, 2020, https://www. standaard.be/cnt/dmf20200211_04844292.
37. Cathy Galle, "Patiënten klagen over privacy e-dossiers: 'Data worden gedeeld zonder toestemming'" [Patients complain about e-file privacy: "Data is shared without permission"], *De Morgen*, December 24, 2019, https://www.demorgen.be/nieuws/patienten-klagen-over-privacy-e-dossiers-data-worden-gedeeld-zonder-toestemming~b0eba24c.
38. Galle, [Patients complain about e-file privacy].
39. Cathy Galle, "Verzekeringsartsen kunnen meekijken in uw medisch dossier" [Insurance doctors have access to your medical file], *De Morgen*, January 23, 2020, https://www.demorgen.be/nieuws/verzekeringsartsen-kunnen-meekijken-in-uw-medisch-dossier~bbcbb6c3.

Chapter 6: The Rise of the Masses

1. Immanuel Kant, Beantwortung to the Frage: Was it Aufklärung? [Answer to the question: What is enlightenment?] *Berlinische Monatsschrift* [Berlin Monthly] (December 1784): 481–94.
2. Hannah Arendt, *The Origins of Totalitarianism* (London: Penguin Books, 1951): 399.
3. Committee P, Annual Report 2019, https://comitep.be/document/jaarverslagen/2019NL_act.pdf.
4. Matthias Verbergt, "Camera's in joodse wijk controleren nu synanggogegangers" [Cameras in Jewish quarter are now used to surveil synagogue-goers], *De Standaard*, March 13, 2021, https://www.standaard.be/cnt/dmf20210312_98151173.
5. Kristof Clerix, "Privacy in corontijden: 'Op de duur zijn we niet meer veraf van Chinese toestanden'" [Privacy in coronavirus times: "In the end, we will not be far away from situations as in China"] *Knack*, February 3, 2021, https://www.knack.be/nieuws/belgie/privacy-in-coronatijden-op-den-duur-zijn-we-niet-meer-veraf-van-chinese-toestanden/article-longread-1695703.html.
6. Melinda Pater, "Je buren verklikken als ze zich niet aan de anderhalve meter houden, het kan" [Denouncing your neighbors if they don't stick to the one and a half meters, it's accepted], *NPO Radio 1*, April 10, 2020, https://www.nporadio1.nl/binnenland/22996-je-buren-verklikken-als-ze-zich-niet-aan-de-anderhalve-meter-houden-het-kan.
7. Collectief van academici [Collective of Academics], "Zonder tegenspraak kan er van wetenschappelijke vooruitgang geen sprake zijn" [There can be no scientific progress without contradiction] *Knack*, April 9, 2021, https://www.knack.be/nieuws/belgie/zonder-tegenspraak-kan-van-wetenschappelijke-vooruitgang-geen-sprake-zijn/article-opinion-1721153.html.
8. International Institute for Democracy and Electoral Assistance Nobelprijswinnaars en wereldleiders: "Coronacrisis bedreigt democratie" [Nobel laureates and world leaders: "Coronavirus crisis threatens democracy"] *Knack*, June 5, 2020, https://www

.knack.be/nieuws/wereld/nobelprijswinnaars-en-wereldleiders-coronacrisis-bedreigt-democratie/article-news-1614169.html.

9. Arendt, *The Origins of Totalitarianism*: 428.
10. Gustave LeBon, *Psychologie des foules* [The crowd: a study of the popular mind] (Paris: Books on Demand, 1895): 17.
11. Le Bon, [The crowd]: 40.
12. Le Bon, [The crowd]: 95–98.
13. Le Bon, [The crowd]: 11–12.
14. Le Bon, [The crowd]: 11.
15. Vivek Murthy, "Work and the Loneliness Epidemic," *Harvard Business Review*, September 26, 2017, https://hbr.org/2017/09/work-and-the-loneliness-epidemic.
16. Liana DesHarnais Bruce, "Loneliness in the United States: A 2018 National Panel Survey of Demographic, Structural, Cognitive, and Behavioral Characteristics," *American Journal of Health Promotion* 33, no. 8 (November 1, 2019): 1123–33, https://doi.org/10.1177/0890117119856551.
17. DesHarnais Bruce, "Loneliness in the United States."
18. Arendt, *The Origins of Totalitarianism*: 415.
19. David Graeber, *Bullshit Jobs* (Amsterdam: Business Contact, 2018).
20. Steve Crabtree, "Worldwide, 13% of Employees Are Engaged at Work," *Gallup World Poll*, October 8, 2013, https://news.gallup.com/poll/165269/worldwide-employees-engaged-work.aspx.
21. DesHarnais Bruce, "Loneliness in the United States."
22. Le Bon, [The crowd]: 25-30.
23. VRT, *De Afspraak* [The Appointment], "Het journal—22 maart 2020," YouTube, 35:23, March 22, 2020, https://www.youtube.com/watch?v=NliNQquAH5M.
24. Freek Willems, "Virologe Vlieghe: 'Mondmasker verplichten creëert bewustzijn dat virus er nog is,' De Block: 'Discussie niet gesloten'" [Virologist Vlieghe: "Mandating a face mask creates awareness that the virus is still there," De Block: "Discussion not closed"] *VRT NWS*, June 28, 2020, https://www.vrt.be/vrtnws/nl/2020/06/28/erika-vlieghe-over-mondmaskers-creeren-awareness-dat-virus-nog.

25. KVE, De Wever (N-VA) haalt uit naar premier en minister Vandenbroucke: "Als je middenstanders te gronde wil richten, dan moet je het zo aanpakken" [De Wever (N-VA) lashes out at Prime Minister and Minister Vandenbroucke: "If you want to destroy retailers, you have to approach it like this"], *HLN*, November 29, 2020, https://www.hln.be/binnenland/de-wever-n-va-haalt-uit-naar-premier-en-minister-vandenbroucke-als-je-middenstanders-te-gronde-wil-richten-dan-moet-je-het-zo-aanpakken~a808260d.
26. Solomon E. Asch, "Effects of Group Pressure upon the Modification and Distortion of Judgment," in H. Guetzkow (ed.), *Groups, Leadership, and Men: Research in Human Relations* (Pittsburgh, PA: Carnegie Press, 1951).
27. Arendt, *The Origins of Totalitarianism*: 148–49.
28. Le Bon, [The crowd]: 21-22.
29. Arendt, *The Origins of Totalitarianism*: 422–24.
30. Torck, L. (2021, April 12). Marc Van Ranst na tragisch weekend: "Onfortuinlijk, maar nul compassie voor feestende jongeren" [Marc Van Ranst after a tragic weekend: "Unfortunate, but zero compassion for young people who party"], *Het Nieuwsblad*, April 12, 2021, https://www.nieuwsblad.be/cnt/dmf20210412_92686979.
31. Le Bon, [The crowd]: 34.
32. Arendt, *The Origins of Totalitarianism*: 455–57.
33. Le Bon, [The crowd]: 26.
34. Félix Julien, *Courants et révolutions de l'athmosphère et de la mer* [Currents and revolutions of the atmosphere and the sea] (Montana: Kessinger Publishing, 1860); Le Bon, [The crowd]: 27.
35. Arendt, *The Origins of Totalitarianism*: 563–64.
36. Paul Aubry, *La contagion du meurtre: étude d'anthropologie criminelle*. [The contagion of murder: a study of criminal anthropology] (Paris: Alcan, 1888).
37. Taine, H. (1893). *Les origines de la France contemporaine: La Revolution* (tome IIV). [The origins of contemporary France: The revolution (volume IIV)] Paris: Hachette.
38. Le Bon, [The crowd]: 32–33.

Chapter 7: The Leaders of the Masses

1. Gustave Le Bon, *Psychologie des foules* [The crowd: a study of the popular mind] (Paris: Books on Demand, 1895): 67.
2. Hannah Arendt, *Eichmann in Jeruzalem* [Eichmann in Jerusalem] (Amsterdam: Olympus, 1963): 195.
3. Arendt, *Eichmann in Jeruzalem*.
4. Arendt, *Eichmann in Jeruzalem*: 211.
5. Arendt, *Eichmann in Jeruzalem*: 220–21.
6. Le Bon, [The crowd]: 33–36.
7. Arendt, *Eichmann in Jeruzalem*: 208–9.
8. Arendt, *Eichmann in Jeruzalem*: 212.
9. Arendt, *Eichmann in Jeruzalem*: 63.
10. Aleksandr Solzhenitsyn, *The Gulag Archipelago* (London: The Harvey Press, 1986).
11. Arendt, *Eichmann in Jeruzalem*: 307.
12. Solzhenitsyn, *The Gulag Archipelago*: 19–38.
13. Hannah Arendt, *The Origins of Totalitarianism* (London: Penguin Books, 1951): 452.
14. Arendt, *The Origins of Totalitarianism*: 575.
15. Arendt, *The Origins of Totalitarianism*: 500.
16. Gunter D'Alquen, *Die SS. Geschichte, Aufgabe, und Organisation der Schutzwaffen der NSDAP* [The SS. History, mission and organization of the Schutztaffeln of the NSDAP] (Berlin: Junker und Dunnhaupt Verlag, 1939).
17. Arendt, *The Origins of Totalitarianism*: 402–3; George Orwell, *Animal Farm* (London: Secker and Warburg, 1945).
18. Arendt, *The Origins of Totalitarianism*: 402–3.
19. Solzhenitsyn, *The Gulag Archipelago*.
20. Arendt, *The Origins of Totalitarianism*: 601.
21. Le Bon, [The crowd]: 67–70.
22. Bruno Bettelheim, *On Dachau and Buchenwald. Nazi Conspiracy*, vol. VII (1946): https://forum.axishistory.com/viewtopic.php?t=68993; David J. Dallin, *From Purge to Coexistence: Essays on Stalin's and Krushchev's Russia* (Chicago: Henri Regnery Company, 1964); and Eugen Kogon, *The Theory and Practice of Hell: The German*

Concentration Camps and the System Behind Them (New York: Farrar, Straus and Giroux, 1956).

23. Arendt, *The Origins of Totalitarianism*: 571, 562, 597, 601.
24. Arendt, *The Origins of Totalitarianism*: 318–19.
25. Arendt, *The Origins of Totalitarianism*: 515–16; Solzhenitsyn, *The Gulag Archipelago*: 120-128.
26. Arendt, *The Origins of Totalitarianism*: 621.
27. Arendt, *The Origins of Totalitarianism*: 619.
28. Arendt, *The Origins of Totalitarianism*: 589.
29. Arendt, *Eichmann in Jeruzalem*: 348.
30. Arendt, *The Origins of Totalitarianism*: 593–94.
31. Orwell, *Animal Farm*.
32. United Nations World Food Program, "2020—Global Report on Food Crises," April 20, 2020, https://www.wfp.org/publications/2020-global-report-food-crises.
33. Arendt, *The Origins of Totalitarianism*: 628.
34. Arendt, *The Origins of Totalitarianism*: 402.
35. Solzhenitsyn, *The Gulag Archipelago*: chapter 2.
36. Arendt, *The Origins of Totalitarianism*: 566; Solzhenitsyn, *The Gulag Archipelago*: 19–38.
37. Solzhenitsyn, *The Gulag Archipelago*: 436–38.
38. Arendt, *The Origins of Totalitarianism*: 446–508.
39. Solzhenitsyn, *The Gulag Archipelago*: 130–31.
40. Solzhenitsyn, *The Gulag Archipelago*: 9.
41. Solzhenitsyn, *The Gulag Archipelago*: 216–17, 221–23.
42. Le Bon, [The crowd]: 13.
43. Solzhenitsyn, *The Gulag Archipelago*: 430.
44. Le Bon, [The crowd]: 13.

Chapter 8: Conspiracy and Ideology

1. Aleksandr Solzhenitsyn, *The Gulag Archipelago* (London: The Harvey Press, 1986).
2. Henri Rollin, *L'apocalypse de notre temps: Les dessous de la propagande Allemande d'après des documents inédits* [The apocalypse of our times: The hidden side of German propaganda according to

unpublished documents] (Paris: Gallimard, 1939), 40.

3. Maurice Joly, *Dialogue aux enfers entre Machiavel et Montesquieu ou la politique de Machiavel au XIX siècle* [Dialogue in hell between Machiavelli and Montesquieu or the politics of Machiavelli in the XIX century] (Brussels: A. Mertens et fils, 1864).
4. Chevalier de Malet, *Recherches politiques et historiques qui prouvent l'existence d'une secte revolutionnaire* [Political and historical research that proves the existence of a revolutionary sect] (Paris: Gide Fils, 1817).
5. Hieronim Zahorowski, *Monita Secreta* (1612), last accessed March 3, 2022, https://ia800503.us.archive.org/32/items/secretamonitasoc00brec/secretamonitasoc00brec.pdf.
6. "Conspiracy," *Wikipedia*, last updated November 4, 2021, https://en.wikipedia.org/wiki/Conspiracy.
7. Gustave Le Bon, *Psychologie des foules.* [The crowd: a study of the popular mind] (Paris: Books on Demand, 1895): 17.
8. Le Bon, [The crowd]: 70–73.
9. Niko Tinbergen, *Inleiding tot de diersociologie* [Introduction to animal sociology] (Gorinchem: Noorduijn and son, 1946).
10. Elias Canetti, *Massa en macht* [Mass and power] (Amsterdam: Athenaeum—Polak & Van Gennep, 2017).
11. Matthias Hides, "Camera's in Joodse wijk controleren nu synagogegangers" [Cameras in Jewish Quarter now monitor synagogue-goers] *De Standaard*, March 13, 2021, https://www.standaard.be/cnt/dmf20210312_98151173.
12. Yuval Noah Harari, *Homo Deus* (London: Vintage, 2015).
13. Zia Khan, "Innovating for a Bold Future," Rockefeller Foundation [blog], October 27, 2020, https://www.rockefellerfoundation.org/blog/innovating-for-a-bold-future.
14. "Event 201," Center for Health Security, last accessed March 3, 2022, https://www.centerforhealthsecurity.org/event201.
15. Klaus Schwab and Thierry Malleret, *COVID-19: The Great Reset* (Agentur Switzerland: World Economic Forum, 2020).
16. Noam Chomsky, *Necessary Illusions: Thought Control in Democratic Societies* (Boston: South End Press, 1989).

17. Departement MOW [MOW Flemish Government Department], "Vlaamse mobiliteitsvisie 2040 : Digi-kosmos" [Flemish mobility vision 2040: Digi-cosmos], YouTube, 1:15, August 11, 2020, https://www.youtube.com/watch?v=mfN3EJMVOQ4.
18. Hannah Arendt, *The Origins of Totalitarianism* (London: Penguin Books, 1951): 541, 569.
19. Hannah Arendt, *Eichmann in Jeruzalem* [Eichmann in Jerusalem] (Amsterdam: Olympus, 1963): 209.
20. Alex Stern, "Sterilization Abuse in State Prisons: Time to Break with California's Long Eugenic Patters," *Huffington Post*, updated September 22, 2013, https://www.huffpost.com/entry/sterilization-california-prisons_b_3631287.
21. Arendt, *The Origins of Totalitarianism*: 470.
22. Arendt, *The Origins of Totalitarianism*: 428.
23. Charles Eisenstein, "The Conspiracy Myth, Charles Eisenstein" [blog], May 2020, https://charleseisenstein.org/essays/the-conspiracy-myth.
24. Arendt, *The Origins of Totalitarianism*: 426.
25. Arendt, *Eichmann in Jeruzalem*: 286.
26. Solzhenitsyn, *The Gulag Archipelago*.

Chapter 9: The Dead versus the Living Universe

1. Pierre-Simon Laplace, *Essai philosophique sur les probabilités* [A philosophical essay on probabilities] (Cambridge: Cambridge University Press, 1795): 4.
2. Bertrand Russell, Letter to Frege (1902), in Jean van Heijenoort (ed.) *From Frege to Gödel: A Source Book in Mathematical Logic, 1879–1931* (Cambridge, Massachusetts: Harvard University Press, 1967): 124–25.
3. Werner Heisenberg, "Über den anschaulichen Inhalt der quantentheoretischen Kinematik und Mechanik" [On the physical content of the quantum theoretical kinematics and mechanics], *Zeitschrift für Physik* [Journal of Physics], 43 (1927): 172–98.
4. James Gleick, *Chaos: Making a New Science* (London: Penguin Books, 1987): 93.
5. Gleick, *Chaos*: 299.

6. Gleick, *Chaos*: 262–67.
7. Gleick, *Chaos*: 43.
8. Hans Meinhardt, *The Algorithmic Beauty of Sea Shells* (Berlin: Springer, 1995).
9. Galileo Galilei, "Il saggiatore/6" [The Taster/6] (1623), last updated April 17, 2011, https://it.wikisource.org/wiki/Il_Saggiatore/6.
10. Edward Lorenz, "Deterministic Nonperiodic flow," *Journal of the Athmospheric Sciences* 20 (March 1963): 130–41.
11. Gleick, *Chaos*: 135.
12. Werner Heisenberg, *Das naturgesetz und die Struktur der Materie* [Natural law and the structure of matter] (Stuttgart: Belser Verlag, 1967).
13. Henri Poincaré, *Science and method* (London: T. Nelson, 1914): https://archive.org/details/sciencemethod00poinuoft/page/n5.
14. Gleick, *Chaos*.

Chapter 10: Matter and Spirit

1. Stephen Hawking and Leonard Mlodinow, *Het grote ontwerp: een nieuwe verklaring van het Universum* [The grand design: A new statement from the universe] (Amsterdam: Bert Bakker, 2010): 75.
2. Hawking and Mlodinow, [The grand design]: 93.
3. Niels Bohr, cited in Karen Barad, *Meeting the Universe Halfway: Quantum Physics and the Entanglement of Matter and Meaning* (London: Duke University Press, 2007): 254.
4. Werner Heisenberg, *Das naturgesetz und die Struktur der Materie* [Natural law and the structure of matter] (Stuttgart: Belser Verlag, 1967).
5. Bertrand Russell, *The Analysis of Mind* (Gutenberg Ebook, 1921): 808.
6. Lionel Feuillet, Henry Dufour, and Jean Pelletier, "Brain of a White-Collar Worker," *The Lancet* 370, no. 9583 (July 1, 2007): 262, https://doi.org/10.1016/S0140-6736(07)61127-1; Roger Lewin, "Is Your Brain Really Necessary?" *Science* 210, no. 4475 (December 12, 1980): 1232–34, https://doi.org/10.1126/science.7434023.
7. Lionel Feuillet, "Brain of a White-Collar Worker."

8. Jan Scholz et al., "Training Induces Changes in White-Matter Architecture," *Nature Neuroscience Online* 12, no. 11 (November 2009): 1370–71, https://doi.org/10.1038/nn.2412; A. M. Clare Kelly and Hugh Garavan, "Human Functional Neuroimaging of Brain Changes Associated with Practice," *Cerebral Cortex* 15, no. 8 (August 2005): 1089–102, https://doi.org/10.1093/cercor/bhi005.
9. Elisabeth Wieduwild et al., "β2-adrenergic Signals Downregulate the Innate Immune Response and Reduce Host Resistance to Viral Infection," *Journal of Experimental Medicine* 217, no. 4 (April 6, 2020): https://doi.org/10.1084/jem.20190554.
10. Anders Prior et al., "The Association between Perceived Stress and Mortality among People with Multimorbidity: A Prospective Population-Based Cohort Study," *American Journal of Epidemiology* 184, no. 3 (August 1, 2016): 199–210, https://doi.org/10.1093/aje/kwv324.
11. Naja Rod Nielsen et al., "Perceived Stress and Cause-Specific Mortality among Men and Women: Results from a Prospective Cohort Study," *American Journal of Epidemiology* 168, no. 5 (September 1, 2008): 481–91, https://doi.org/10.1093/aje/kwn157.
12. H. F. Ellenberger, *The Discovery of the Unconscious* (New York: Basic Books, 1970).
13. Christine Watremez and Fabienne Roelants, "Hypnose in de anesthesie" [Hypnosis in anesthesia], *Bloedvaten, Hart, Longen* [Blood Vessels, Heart, Lungs] 15, no. 1 (2010): 30–34.
14. Arthur Shapiro, *The Powerful Placebo* (Baltimore: The Johns Hopkins University Press, 1997); Bruce E. Wampold, "The Placebo Is Powerful: Estimating Placebo Effects in Psychotherapy and Medicine from Randomized Clinical Trials," *Journal of Clinical Psychology* 61, no. 7 (July 2005): 835–54, https://doi.org/10.1002/jclp.20129.
15. A. Hróbjartsson and P. C. Gøtzsche, "Is the Placebo Powerless? An Analysis of Clinical Trials Comparing Placebo with No Treatment," *New England Journal of Medicine* 344, no. 21 (May 24, 2001): 1594–1602, https://doi.org/10.1056/NEJM200105243442106.

16. Robert A. Hahn, "The Nocebo Phenomenon: Concept, Evidence, and Implications for Public Health," *Preventive Medicine* 26 (1997): 60711.
17. L. Harrison Matthews, "Visual Stimulation and Ovulation in Pigeons," *Proceedings of the Royal Society, Series B (Biological Sciences)* 126, no. 845 (February 3, 1939): 557–60.
18. Rémy Chauvin, "Contribution à l›étude physiologique du criquet pèlerin et du déterminisme des phénomènes grégaires" [Contribution to the physiological study of the desert locust and the determinism of gregarious phenomena], *Bulletin de la Société entomologique de France* 85, no. 7–8 (1980): 133–272, https://doi.org/10.3406/bsef.1980.18263.
19. Marcel Mauss, *Essai sur le don: Forme et raison de l'échange dans les sociétiés primitives* [Essay on the gift: Form and purpose of exchange in primitive societies], February 2002), https://anthropomada.com/bibliotheque/Marcel-MAUSS-Essai-sur-le-don.pdf.
20. Claude Lévi-Strauss, "L'efficacité symbolique" [Symbolic efficiency], *Revue de l'histoire des religions* 135, no. 1 (1949): 5–27, https://doi.org/10.3406/rhr.1949.5632.
21. Aleksandr Solzhenitsyn, *The Gulag Archipelago* (London: The Harvey Press, 1986): 318–19.

Chapter 11: Science and Truth

1. Bertrand Russell, *The Impact of Science on Society* (1953; London: Routledge, 2013).
2. Jacob Fox, "Essay: COVID-19, Utopianism, and the Reimagination of Society," *Collateral Global*, October 17, 2021, https://collateralglobal.org/article/covid-19-utopianism-and-the-reimagination-of-society.
3. Georg W. F. Hegel, *Vorlesungen über die Philosophie der Religion* [Lectures on the philosophy of religion] (1821; Hamburg: Felix Meiner Verlag, 1993).
4. Niels Bohr, cited in Steve Giles, *Theorizing Modernism: Essays in Critical Theory* (1920; London: Routledge, 1993).

5. Max Planck, *Scientific Autobiography and Other Papers*, trans. Frank Gaynor (New York Philosophical Library: 1949; Westport, CT: Greenwood Press, 1971): 184.
6. Ken Wilber, *Quantum Questions: Mystical Writings of the World's Greatest Physicist* (Boulder: Shambala, 1998): 16.
7. René Thom, *Predire n'est pas expliquer* [To predict is not to explain], trans. Roy Lisker (Champs Sciences, Editions Eshel, IHES edition, 2010): 92.
8. Masaaki Hatsumi, *Essence of Ninjutsu: The Nine Traditions* (Chicago: Contemporary Books, 1988).
9. Max Jacob, *Le cornet a des* [Dice box] (Paris: Jourde and Allard, 1917).
10. Michel Foucault, *De moed tot waarheid* [The courage of truth] (Amsterdam: Boom, 1983).
11. Foucault, [The courage to truth]: 45.

INDEX

Note: Page numbers in *italics* refer to figures. Page numbers followed by t refer to tables.

D

K

L

M

N

ABOUT THE AUTHOR

Mattias Desmet is recognized as the world's leading expert on the theory of mass formation as it applies to the COVID-19 pandemic. He is a professor of clinical psychology in the Department of Psychology and Educational Sciences at Ghent University (Belgium) and a practicing psychoanalytic psychotherapist. His work has been discussed widely in the media, including on *The Joe Rogan Experience* and in *Forbes*, *The New York Post*, Salon.com, and *Fox News*, among hundreds of other outlets. His interviews have been viewed by millions of people around the world. His previous books include *The Pursuit of Objectivity in Psychology* and *Lacan's Logic of Subjectivity: A Walk on the Graph of Desire*. Desmet is the author of over one hundred peer-reviewed academic papers. In 2018 he received the Evidence-Based Psychoanalytic Case Study Prize of the Association for Psychoanalytic Psychotherapy, and in 2019 he received the Wim Trijsburg Prize of the Dutch Association of Psychotherapy.